Ordinary Differential Equations
Practice, Techniques, and Theory

Alice Gorguis
North Park University

January'2014

Copyright © 2014 by Alice Gorguis.

ISBN:	Softcover	978-1-4990-6036-2
	eBook	978-1-4990-6037-9

All rights reserved. No part of this book may be reproduced or transmitted in any form or by any means, electronic or mechanical, including photocopying, recording, or by any information storage and retrieval system, without permission in writing from the copyright owner.

Any people depicted in stock imagery provided by Thinkstock are models, and such images are being used for illustrative purposes only.
Certain stock imagery © Thinkstock.

This book was printed in the United States of America.

Rev. date: 08/14/2014

To order additional copies of this book, contact:
Xlibris LLC
1-888-795-4274
www.Xlibris.com
Orders@Xlibris.com
653508

Dedication:

This book is dedicated to the memory of my beloved mother, who was always there supporting me through my bleak days with her endless love that inspired me to do the best.

Alice Gorguis.

Preface:

This is a text-book for one semester course in differential equations designed for students who have completed the ordinary course in elementary calculus. I hope that this book can enable the student to learn enough examples, and methods in differential equations. It has been the purpose to limit the amount of material to what can readily be covered in the class room for a semester of four hours per week, and to select those topics for discussion which the student might well know who expects to teach differential equations or to continue his/her study of differential equations.

In this course differentiation and integrating will be used extensively, and if the student has forgot differentiation and integration studied in Calculus, this course will help him/her to remember it and learn more, so after this course students will definitely master both differentiation and integration, and be able to apply it to the real world problems.

My goal of writing this book is to help the students to learn the style of thinking, and learn the basic ideas and techniques of differentiation and integration for all type of differential equations both linear, non linear, homogeneous and non homogeneous. I include some exercises and examples that help to connect the abstract concepts being discussed to the student's own understanding of more concrete ideas.

Methods of Teaching: A typical class will be a mixture of lecture, discussion, and problem solving followed by group work on problem solving in the class.

Students will be given homework assignments and problems to work on. Also , projects using Mathematica will be given to students to work on as a meaning of using technology to solve differential equations.

A. Gorguis
January' 2014

Contents

1   **11**

    1.1   Historical Notes 11
    1.2   Classifications of Differential Equations . . 12
    1.3   Definitions and Terminology 15
    1.4   Type of Solutions 19
    1.5   The Initial Value Problem 22
    1.6   Existence and Uniqueness of IVP 23
    1.7   Technical Writing 25
    1.8   Review Exercise 25
    1.9   Review Exercise Solutions 27
    1.10 Chapter-1 Assignment 30

2   **33**

    2.1   The Falling Object 33
    2.2   Separable Equations 36
    2.3   Exact Equations 41
    2.4   Linear Equations 46
    2.5   Homogeneous Equations 50
    2.6   Non-Linear equations and
            Solutions 54
    2.7   Applications of
            First Order Equations 60
    2.8   Technical Writing 65
    2.9   Review Exercises 66

	2.10	Review Exercise Solutions	68
	2.11	Chapter-2 Assignment	81

3   83

    3.1   Introduction 83
    3.2   The Simple Pendulum 83
    3.3   The Wronskian for 2nd ODE 87
    3.4   Finding a Second Solution
           from a Given One 93
    3.5   Method of Reduction
           of Order(2) 97
    3.6   Homogeneous Equations with 99
    3.7   Non-Homogeneous Eq's
           with Constant Coefficients 104
    3.8   A Special Case of
           Variable Coefficients 116
    3.9   Technical Writing 125
    3.10 Review Exercise 125
    3.11 Review Exercise Solutions 128
    3.12 Chapter-3 Assignment 139

4   141

    4.1   Introduction 141
    4.2   Definition of Laplace
           Transforms 142
    4.3   Properties of Laplace
           Transform 146
    4.4   Inverse Laplace Transform 147
    4.5   Solving Initial Value
           Problems. 148
    4.6   Decomposition of
           Partial Fraction: 151
    4.7   Laplace for
           Unit-Step-Functions: 160

- 4.8 Applications of Laplace Transform 174
- 4.9 Technical Writing 179
- 4.10 Review Exercise 180
- 4.11 Review exercise Solutions 180
- 4.12 Chapter-4 Assignment 183

5   185

- 5.1 Differentiation, Integration of Power Series 189
- 5.2 Initial Value Problems of Power Series 190
- 5.3 Technical Writing 202
- 5.4 Review Exercise 202
- 5.5 Review Exercise Solutions 202
- 5.6 Chapter-5 Assignment 207

6   Review from Calculus   209

- 6.1 The Chain Rule 213
- 6.2 Important Formula's 219

INDEX   222

Chapter 1

Introduction

Mathematical applications involving rates of change lead to equations containing derivatives, such equations are called differential equations. Students first encounter with differential equations was in calculus course when anti derivatives (indefinite integrals) were studied.

1.1 Historical Notes

Differential equations is one of the oldest subject in modern mathematics, its origin dates back to Newton's classification of first order ordinary differential equations in 1727 into three classes. Newton used differential equation in his study of planetary motion and also in his consideration of optics. Newton assumed that the solution of differential equation could be expressed as an infinite series, and he successfully defined the coefficients in a manner similar to todays techniques, but he did not consider the convergence of the series.

Newton introduced the notation $y^\bullet$ for the derivative of y with respect to the independent variable.

Lebiniz (1646-1716) invented the differential notation dy, and he used the term "differential equations" to denote a relationship between two differential dx and dy and also, he invented the symbol $\int$ for integration. In 1691 Leibniz discovered the method of separation of variables and then he reduced the linear homogeneous first order differential equations to quadratures.

Jean Bernoulli Used the word Integral in 1690, and he successfully studied the general homogeneous linear differential equations with constant coefficient. Leonard Euler (1707-1783) introduced the method of variation of parameters. Cauchy (1789 - 1857) developed the existence and uniqueness theorem for first order ordinary differential equations, and extended it into n-dependent variables. Rudolf Lipschitz (1832 - 1903) generalized Cauchy's existence and uniqueness theorem in 1876.

Picard (1856-1941) improved Cauchy's and Lipschitz's theorem by introducing the method of successive approximations in 1893.

1.2 Classifications of Differential Equations

By differential equations (DE) we mean any equation that involves derivatives or differentials of a function or functions. The order of a differential equation is the largest positive integer n, e.g if a differential equation is written as a polynomial then the highest power

derivative is called the degree of the polynomial.
The differential equations are divided into 2-categories:

- Ordinary differential equations (ODE).
- Partial differential equations (PDE).

This classification is based on the unknown function in the differential equation. If the unknown depends on one-independent variable only s.a. $y' = f(x)$, then the differential equation is called ordinary differential equation, written as $y' = dy/dx$, and symbolized as ODE. If the unknown function depends on more than one independent variable s.a. $y' = f(x, y)$, then the differential equation is called partial differential equation, written as $y' = \partial y / \partial x$ and symbolized as PDE.

Examples:
a. The following equations are ordinary differential equations:

$$\frac{d^2y}{dx^2} + \frac{dy}{dx} + y = 0. \tag{1.1}$$

$$8t\frac{dx}{dt} + 2cx = 0. \ c \ is \ a \ constant. \tag{1.2}$$

$$y'' + y = 0. \tag{1.3}$$

b. The followings equations are partial differential equations:

$$\frac{\partial^2 u}{\partial x^2} + \frac{\partial^2 u}{\partial y^2} = 0. \tag{1.4}$$

Where $0 < x < a$, and $0 < y < b$, and the equation is known as the Laplacian.

$$\frac{\partial u}{\partial t} = k\frac{\partial^2 u}{\partial x^2}. \tag{1.5}$$

This is known as one-dimensional Heat - equation, where k is constant, and u is differentiable once with respect to time t, and twice with respect to space x. And,

$$\frac{\partial^2 u}{\partial t^2} = k\frac{\partial^2 u}{\partial x^2}. \qquad (1.6)$$

Which is one-dimensional Wave-equation with constant k, and u is twice differentiable with respect to both time t and space x.

Both ODE and PDE are subdivided into two classes of differential equations:
- Linear differential equations.
- Non-Linear differential equations.

The Linear Differential equation in general (nth-order)has the following form:

$$a_n(x)\frac{d^n y}{dx^n} + a_{n-1}(x)\frac{d^{n-1} y}{dx^{n-1}} + \ldots + a_1(x)\frac{dy}{dx} + a_0(x)y = f(x). \qquad (1.7)$$

Any Ordinary Differential equation that can not be written in the form (1.7) is considered to be non-linear ODE.

Also, both ODE and PDE are subdivided into two sub-classes of differential equations:
- Homogeneous differential equations.
- Non-Homogeneous differential equations.

Equation(1.7) is a Non-Homogeneous equation because of the function $f(x)$.

When $f(x) = 0$ in equation 91.7), then the function is considered a Homogeneous function,

$$a_n(x)\frac{d^n y}{dx^n} + a_{n-1}(x)\frac{d^{n-1}y}{dx^{n-1}} + \ldots + a_1(x)\frac{dy}{dx} + a_0(x)y = 0. \tag{1.8}$$

1.3 Definitions and Terminology

1. An equation containing the derivative of one or more dependent variables, with respect to one or more independent variables, is said to be a differential equation.
2. If the equation contains only ordinary derivatives of one or more dependent variables with respect to a single independent variable, it is then said to be ODE.
3. If the equation involves partial derivatives of one or more dependent variables of two or more independent variables , it is called a partial differential equation.
4. The order of the higher derivative in a DE is called the order or degree of the equation or degree, e.g,

$$\frac{d^3 y}{dx^3} + (\frac{dy}{dx})^2 - y = 2x, \tag{1.9}$$

is a 3rd ODE, or ordinary differential equation of degree three and,

$$\frac{\partial^4 u}{\partial t^4} + \frac{\partial u}{\partial t} = 0, \tag{1.10}$$

is a 4th PDE, or partial differential equation of degree 4.
5. A differential equation is said to be Homogeneous when $f(x) = 0$,or is in the form of equation (1.8), and non-homogeneous if $f(x) \neq 0$ or in the form of equation (1.7).

Symbols

Symbol	Description
DE	differential equations
ODE	ordinary differential equations
PDE	partial differential equations
dy/dx	first order ODE with respect to x
d^2y/dx^2	second order ODE with respect to x
$\partial y/\partial x$	first order PDE with respect to x
$\partial^2 y/\partial x^2$	second order PDE with respect to x
dy/dx	can be written as y'
$\partial y/\partial x$	can be written as y_x

Table(1)

Linear Differential Equations

The linear ordinary differential equation is any equation that resembles equation (1.7),

$$a_n(x)\frac{d^n y}{dx^n} + a_{n-1}(x)\frac{d^{n-1}y}{dx^{n-1}} + \dots + a_1(x)\frac{dy}{dx} + a_0(x)y = f(x).$$

with the following descriptions:
y is the dependent function, x is the independent function, and all the coefficients : a_n, a_{n-1}, ... a_1, a_0 and also the Non-homogeneous function f are functions of the independent variable x. An equation that does not satisfy any of these descriptions is called non-linear equation. Thus to classify the equation one must think of comparing it with the form on equation 1.7.

Examples

Give a full Classification to the following equations (Hint: the best way to classify each equation is to compare it with equation (1.7):

1.3. DEFINITIONS AND TERMINOLOGY

$$x\frac{dy}{dx} + y = 0. \tag{1.11}$$

This is first order *Linear* ordinary differential equation because $n = 1$, homogeneous because $f(x) = 0$, y is the dependent variable, x is the independent variable, and the leading coefficient is a function of x.

$$\frac{d^2x}{dt^2} + \frac{dx}{dt} - x = 2t. \tag{1.12}$$

This is second order *Linear* ordinary differential equation because $n = 2$, Non-homogeneous because $f(t) = 2t$, x is the dependent variable, t is the independent variable, and the leading coefficient is 1.

$$x^3 y''' + x^2 y'' + xy' + y = \cos x \tag{1.13}$$

This is a third order *Linear* ordinary differential equation because $n = ''' = 3$, Non-homogeneous because $f(x) = \cos x$, y is the dependent variable, x is the independent variable, and the leading coefficients (x^3, x^2, x) are all functions of the independent variable x.

$$y\frac{dy}{dx} + y = 0. \tag{1.14}$$

This is a first order because $n = 1$, but $Non - Linear$ ordinary differential equation because the leading coefficient is y which is not a function of the independent variable x. The function is homogeneous because $f(x) = 0$, y is the dependent variable, and x is the independent variable.

$$y'' + y^2 = 0. \tag{1.15}$$

This is a second order because $n =''= 2$, but *Nonlinear* ordinary differential equation because the second leading coefficient is y which is not a function of the independent variable x. The function is homogeneous because $f(x) = 0$, y is the dependent variable, and x is the independent variable.

General Remark: Sometimes it is difficult to recognize the differential equation as a liner in one function or the other, so one has to be careful in finding out by manipulating the equation in writing it in different form to see it clear. The following example illustrates that:

Example - 1:
Determine if the differential equation :

$$(x^3 - 2)dt + tdx = 0. \qquad (1.16)$$

Is linear in x or in t.

Solution:
To rewrite the equation in a linear form as follow:

$$(x^3 - 2)dt + tdx = 0. \qquad (1.17)$$

$$t\frac{dx}{dt} = -(x^3 - 2). \qquad (1.18)$$

$$t\frac{dx}{dt} + x^3 = 2. \qquad (1.19)$$

Equation (1.19) is non-linear in x, if we multiply equation (1.18) by dt/dx, and rearranging, we get,

$$(x^3 - 2)\frac{dt}{dx} + t = 0. \qquad (1.20)$$

which is a linear equation in t.

1.4 Type of Solutions

To solve the n-th order Homogeneous equation of the form (1.8), first we will rewrite the equation in the following form:

$$F(\frac{d^n y}{dx^n}, \frac{d^{n-1} y}{dx^{n-1}}, ..., \frac{dy}{dx}, y, x) = 0 \qquad (1.21)$$

Where, y is the dependent, and x is the independent functions. We can assume that x lies in an open interval (a, b), or close interval $[a, b]$, or half open intervals $[(a, b], or [a, b)]$. Then, we try to solve the above equation for the highest order of $d^n y/dx^n$ as,

$$\frac{d^n y}{dx^n} = F(\frac{d^{n-1} y}{dx^{n-1}}, ..., \frac{dy}{dx}, y, x) \qquad (1.22)$$

Now we will choose a function say $u(x)$ and substitute it for y in (1.22). If $u(x)$ satisfies the equation for all x in the given interval, then the function $u(x)$ is an **Explicit Solution** to the equation in (1.22).

Definition : Explicit Solution
A function $u(x)$ that satisfies a DE such as (1.22) for all x in the interval, is called the **explicit solution**.

Example – 2 :
Show that $u(x) = x^3$ is an explicit solution to the given differential equation on the interval $(-\infty, \infty)$

$$x \frac{dy}{dx} = 3y. \qquad (1.23)$$

Solution :
First, we solve for the highest order or $\frac{dy}{dx}$,

$$\frac{dy}{dx} = 3\frac{y}{x}. \qquad (1.24)$$

Then substitute $u(x)$ for y gives,

$$\frac{du(x)}{dx} = 3\frac{u(x)}{x} \qquad (1.25)$$

$$\frac{d}{dx}(x^3) = 3\frac{x^3}{x}. \qquad (1.26)$$

$$3x^2 = 3x^2 \qquad (1.27)$$

This proves that $u(x) = x^3$ is the **Explicit Solution** to the equation in (1.23).

Definition : Implicit Solution
A solution $f(x, y) = 0$ is said to be implicit solution to equation (1.22) on the interval I if it defines one or more explicit solutions in I.

Example – 3 :
Show that,

$$x^2 - t^2 + 4 = 0. \qquad (1.28)$$

Is an implicit solution to the following differential equation,

$$\frac{dx}{dt} = \frac{t}{x}, \qquad (1.29)$$

on the interval $(2, \infty)$.
Solution :
From the given relation in (1.28) we can solve for x first,

$$x = \pm\sqrt{t^2 - 4}. \qquad (1.30)$$

Let $u(t) = x = +\sqrt{t^2 - 4}$, then differentiating with respect to t gives,

$$u'(t) = \frac{dx}{dt} = \frac{1}{2}(t^2 - 4)^{-\frac{1}{2}}(2t) = \frac{t}{\sqrt{t^2 - 4}}. \qquad (1.31)$$

1.4. TYPE OF SOLUTIONS

And substituting into equation (1.29) gives,

$$\frac{t}{\sqrt{t^2-4}} = \frac{t}{\sqrt{t^2-4}}. \tag{1.32}$$

Notice both $u(t)$, and $u'(t)$ are defined in the interval $(2, \infty)$, and the solution is valid for all t in the interval. In the same way we can prove the validity of the negative solution.

General Remark :
To test if a function is a solution for a given differential equation or not, we differentiate the function as many times up to the order of the differential equation, then substitute the function and its derivatives into the differential equation, then prove the right-hand-side of the differential equation is equal to the left-hand-side. This is illustrated it in the following example:

Example – 4 :
Show that the function, $u(t) = 4\,cos2t - 8\,sin2t$ is a solution for the differential equation:

$$\frac{d^2 u(t)}{dt^2} + 4u(t) = 0. \tag{1.33}$$

Solution :
Since the differential equation is of order 2, then we will differentiate the function $u(t)$ twice, then substitute it back:

$$u(t) = 4\,cos2t - 8\,sin2t.$$
$$\frac{du(t)}{dt} = -8\,sin2t - 16\,cos2t.$$
$$\frac{d^2(t)}{dt^2} = -16\,cos2t + 32\,sin2t.$$

Substituting all back into the equation (1.33) and simplifying, we can prove the L.H.S = R.H.S = 0.

1.5 The Initial Value Problem

For the differential equation (DE) of the form:

$$F(\frac{d^n y}{dx^n}, \frac{d^{n-1} y}{dx^{n-1}}, ..., \frac{dy}{dx}, y, x) = 0 \qquad (1.34)$$

The solution in an interval I that satisfies, at x_0 the n-initial conditions,

$$y(x_0) = y_0,$$
$$y(x_1) = y_1,$$
$$\vdots$$
$$y(x_{n-1}) = y_{n-1},$$

where $x_0 \in I$, and $y_0, y_1, .., y_{n-1}$ are given constants, the problem then is called **Initial Value Problem** and symbolized as IVP.

Example – 5 :
Show that $u(t) = Ae^{2t} + Be^{-t}$ is a solution to the DE,

$$y'' - y' - 2y = 0. \qquad (1.35)$$

Find the constants A, and B so that the following initial conditions are satisfied,

$$y(0) = 2,$$
$$y'(0) = 1.$$

1.6. EXISTENCE AND UNIQUENESS OF IVP

Solution :
Using the given solution and differentiating it twice,

$$u(t) = Ae^{2t} + Be^{-t}.$$
$$u'(t) = 2Ae^{2t} - Be^{-t}.$$
$$u''(t) = 4Ae^{2t} + Be^{-t}.$$

Applying the initial conditions we get,

$$u(0) = 2 = Ae^0 + Be^0 = A + B.$$
$$u'(0) = 1 = 2Ae^0 + Be^0 = 2A - B.$$

Then solving the system of two equations for the constants A, and B gives: $A = 1$, and $B = 1$, then the solution of the IVP is,

$$u(t) = e^{2t} + e^{-t}.$$

1.6 Existence and Uniqueness of IVP

In seeking methods of finding the solution to differential equations, w assume that the solution exists and is unique. Since the differential equation is only a model of a physical problem, it is possible that the model is a bad one and has no solution, so instead of wasting time and money, it is wiser to test for existence of the solution and its uniqueness. Similarly for uniqueness theorem, if we know the problem has a unique solution, then once we have found one solution, we are done.

Theorem : For the given IVP,

$$\frac{dy}{dx} = f(x, y), \qquad (1.36)$$

with initial condition $y(x_0) = y_0$.

If $f(x, y)$, and $f'(x, y)$ are continuous function in the domain $\mathbf{R}$ where, $\mathbf{R} = \{(x, y) : a \leq x \leq b, c \leq y \leq d\}$, with point (x_0, y_0) in the domain, then the IVP has a unique solution in the interval $x_0 - h < x < x_0 + h$, where h is a positive number.

Example – 6 :
Prove that the solution is a unique one for the IVP,

$$\frac{dy}{dx} = xy^2 - x^3, \qquad (1.37)$$

for the initial condition $y(1) = 4$.
Solution :
Using the above theorem,

$$f(x, y) = xy^2 - x^3, is\ continuous\ on\ (1, 4)$$
$$and\ \frac{\partial f}{\partial y} = 2xy, also\ continuous\ on\ (1, 4)$$

Since both of the functions are continuous, then the IVP has a unique solution in the interval $(1, 4)$, and its solution falls within the interval $(1 - h, 1 + h)$, where h is a positive number.

1.7 Technical Writing

This section is to help the students to learn how to write Mathematically and apply the concepts of differential equations to the real world and from point of views of different disciplines.

Exercise Problem: Discus situations from the following disciplines: Physics, Astronomy, Biology, and Ecology, where differential equations are used to solve a problem. Describe the situation as: Explicit, implicit, linear or non-linear.

1.8 Review Exercise

In problems 1 through 6, classify the differential equations as: ODE, PDE, Linear, Non-Linear, Homogeneous, Non-Homogeneous, and give the order of the equation, and state the dependent and independent variable:

1. $\dfrac{d^3u}{dt^3} + 4\dfrac{d^2u}{dt^2} - 3\dfrac{du}{dt} + 9u = 1/2 cos 2t$.

2. $3x\dfrac{d^2y}{dx^2} - 5\dfrac{dy}{dx} + xy = x^3$.

3. $9\dfrac{d^5y}{dx^5} = x(1-x^2)$.

4. $e^{x^2}\dfrac{dy}{dx} + 3e^{x^2}y = xe^{x^2}$.

5. $x'' = k(4-x), k = constant$.

6. $$\frac{dx}{dt} = \frac{x(1-t)}{t(1+t)}.\qquad(1.38)$$

In problems 7 − 9 determine whether the given function is a solution to the given differential equation:

7. $\quad x = e^{2t} - 3e^{-t}; \quad x'' - x' - 2x = 0.\qquad(1.39)$

8. $\quad u = cost + t^2; \quad u'' + u = t^2 + 2.\qquad(1.40)$

9. $\quad y = 3cos2x + 6 - 2x; \quad y'' + 3y = 4e^{-x}.\qquad(1.41)$

10. Show that:
$$u(x) = e^{-x^2} \int_0^x s^2 e^{s^2} ds.$$

Is the solution for the differential equation :
$u'(x) + 2xu = x^2.$

11. Find the explicit solution for the IVP:
$$\frac{dy}{dx} = 3x^3 e^{-2y}; \quad y(0) = 2.$$

12. Find the Implicit solution for the differential equation:
$$\frac{dy}{dx} = \frac{y^3}{x}; \quad x \neq 0.$$

In problems 13 − 15 determine whether the IVP has a unique solution in the given interval:

13. $\qquad y' = \frac{x}{y}; \quad y(2) = 0.$

14. $\qquad y' = x^2 - y^2; \quad y(1) = 6.$

15. $\qquad y' = 1 + y^2, \text{ at } (x_0, y_0).$

16. Show if the basic uniqueness theorem applies to the IVP:
$$y'(t) = y^{2/3} \quad y(0) = 0.$$

1.9 Review Exercise Solutions

1. The differential equation is linear ODE, 3rd order, non-homogeneous with $f(t) = 1/2cos(2t)$, dependent variable u, and t as independent variable.

2. The differential equation is linear ODE, 2nd order, non-homogeneous where $f(x) = x^3$, with dependent variable y, and x as independent variable.

3. Comparing with equation (1.7), the differential equation is linear ODE, 5th order, non-homogeneous where $f(x) = x(1 - x^2)$, with dependent variable y, and x as independent variable.

4. The differential equation is linear ODE, 3rd order, non-homogeneous where $f(x) = xe^{x^2}$, with dependent variable y, and x as independent variable.

5. Rearranging the equation as: $x'' + kx = 4k$ proves to be: linear ODE, 2nd order, non-homogeneous where $f(t) = 4k(constant)$, with dependent variable x, and t or any choice of variable, as independent variable.

6. Rearranging the DE,
$$\frac{dx}{dt} = \frac{x(1-t)}{t(1+t)}.$$
$$\frac{dx}{dt} - \frac{(1-t)}{t(1+t)}x = 0.$$
Proves to be linear ODE, 1st order homogeneous with $f(t) = 0$.

7. yes.

8. yes.
9. No.

10. $u(x) = e^{-x^2} \int_0^x s^2 e^{s^2} ds.$

$u'(x) = e^{-x^2}(x^2 e^{x^2}) - 2xe^{-x^2} \int_0^x s^2 e^{s^2} ds.$

Substituting in the given equation:

$u'(x) + 2xu(x) = x^2,$ we get,

$e^{-x^2}(x^2 e^{x^2}) - 2xe^{-x^2} \int_0^x s^2 e^{s^2} ds + 2xe^{-x^2} \int_0^x s^2 e^{s^2} ds = x^2$

Since the right side equal the left side, then the $u(x)$ is the solution to the DE.

11. Separating the differential equation as shown:

$$\frac{dy}{dx} = 3x^3 e^{-2y}.$$
$$e^{2y} dy = 3x^3 dx.$$
Then integrating both sides, $\int e^{2y} dy = 3 \int x^3 dx.$
$$1/2 e^{2y} = 3/4 x^4 + c.$$

Substituting the initial values $y(0) = 2$ to find the value of the constant c gives, $c = 1/2e^4$. Thus the explicit solution is:

$$1/2 e^{2y} = 3/4 x^4 + 1/2 e^4.$$
$$Or\ y(x) = \frac{1}{2} ln(\frac{3}{2} x^4 + e^4).$$

1.9. REVIEW EXERCISE SOLUTIONS

12. Separating the DE, then integrating,

$$\frac{dy}{dx} = \frac{y^3}{x}; \quad x \neq 0.$$
$$\frac{dy}{y^3} = \frac{dx}{x}.$$
$$\int \frac{dy}{y^3} = \int \frac{dx}{x}.$$
$$-\frac{1}{2}y^{-2} = \ln x + c.$$

The final equation is the implicit one.

13. $f(x, y) = \frac{x}{y}$, and $\frac{\partial f}{\partial y} = \frac{-x}{y^2}$. Since both f, and f' are continuous in the given interval, then the solution is unique.

14. $f(x, y) = x^2 - y^2$, and $\frac{\partial f}{\partial y} = -2y$. Since both f, and f' are continuous in the given interval, then the solution is unique.

15. $f(x, y) = 1 + y^2$, and $\frac{\partial f}{\partial y} = 2y$. Since both f, and f' are continuous, then the solution is unique.

16. $f(x, y) = y^{\frac{2}{3}}$, and $\frac{\partial f}{\partial y} = 2/3 y^{-1/3} = \frac{2}{3y^{1/3}}$. f' is not continuous in the domain, then the solution is not unique.

1.10 Chapter-1 Assignment

In problems 1 - 6 classify whether the differential equation is linear or nonlinear:

1. $\dfrac{dy}{dx} = 1 - x.$

2. $\dfrac{d^2y}{dx^2} + \dfrac{dy}{dx} = \cos y.$

3. $\dfrac{d^y}{dx^2} = 1 + (\dfrac{dy}{dx})^2.$

4. $y^2 dx + x dy = 0.$

5. $y'' + y' + 9y = \sin x.$

6. $(x^2 - 1)dy + y dx = 0.$

In problems 7 - 12 determine if the given function is a solution to the given DE:

7. $\dfrac{d^2y}{dt^2} + y = 0;\ y(t) = \sin t.$

8. $= t - y;\ y(t) = t.$

9. $xy'' + -y' + 4x^2 y = 0;\ y(x) = 3\cos x^2 - 4\sin x^2.$

10. $y'' - y' - y = 0;\ y(x) = c_1 e^{\frac{1+\sqrt{5}}{2}} + c_2 e^{\frac{1-\sqrt{5}}{2}}.$

1.10. CHAPTER-1 ASSIGNMENT

11. $y' + y = 0;\quad y(x) = e^{-rx}$.

12. $\dfrac{dy}{dx} + x^2 y = 0;\quad y(x) = e^{-x^3/3}$.

In problems 13 - 14 determine whether the given relation is the implicit solution to the given differential equations:

13. $t^2 + x^2 = 2;\quad \dfrac{dx}{dt} = \dfrac{t}{x}$.

14. $lin\left|\dfrac{x}{1-x}\right| = t + c;\quad x' = x(1-x)$.

In problems 15 - 18 find an explicit solution to the given differential equations:

15. $x^2 \dfrac{dy}{dx} = y$.

16. $\dfrac{dy}{dx} = e^y$.

17. $\dfrac{dy}{dx} = y + 2$.

18. $y' = \dfrac{dy}{dt} = t^3$.

In problems 19 - 23 Find a general solution to the following equations:

19, $u' + 3u = 0$.

20. $y' - \sqrt{1 + 2xy} = 0$.

21. $y' - ln(x)y = 0$.

22. $y' = 3yx^2$.

23. $y' = sin(2x)$.

In problems 24 - 25 Solve the initial value problem:

24. $\dfrac{dy}{dx} = x^2 e^y; \quad y(0) = 1$.

25. $u' = u^2 + 2; \quad u(0) = 5$.

In problems 26 - 27 Determine whether the given initial value
problem has a unique solution:

26. $y' = \dfrac{x}{y}; \quad y(0) = 0$.

27. $y\dfrac{dy}{dx} = x^2; \quad y(1) = 0$.

Chapter 2

First Order Differential equation

Differential equations are extremely important in solving problems in all disciplines in the physical sciences. Students are required to be able to derive the necessary differential equations and solve them. We shall consider some physical situation that lead to differential equations and present the theoretical means for obtaining their solutions.

2.1 The Falling Object

We will start with the motion of a falling object, and the question stated as:
Consider an object falling toward the earth. Assuming that the only forces acting on the object are the gravity, and air resistance. Determine the velocity of the object as a function of time.

To solve this problem we will use Newton's second law, which states that: The force is equal to the mass times acceleration, which can be expressed in the following equation:

$$F = ma \qquad (2.1)$$

Where, (F) is the total forces acting on the falling object, (m) is the mass of the object(constant), and (a) is the acceleration of the object. Since the acceleration is defined as the rate of change of velocity in time, then it can be written as:

$$a = \frac{dv}{dt}. \qquad (2.2)$$

Then equation (2.1) can be written as,

$$F = m\frac{dv}{dt} \qquad (2.3)$$

But the force F is the resultant of two vector forces acting vertically on the object, one of them is the weight of the object $w = mg$ downward vector, where g is the gravity, and the other one is the air resistance force(vector) opposite to the direction of falling and represented by $(-kv)$, taking the downward direction to be positive, then the resultant force on the object is :

$$F = mg - kv, \qquad (2.4)$$

whee, k is the air resistance constant (positive). Assuming that the object starts with initial velocity v_0, then the problem turned to be a first order initial value problem IVP as follows:

$$m\frac{dv}{dt} = mg - kv, \ v(0) = v_0. \qquad (2.5)$$

2.1. THE FALLING OBJECT 35

The question here is: How to solve the initial value problem (2.5)?
After studying the method of solving similar type of problems, we will get back and solve this problem.

There are many more examples of applications of first order differential equations in Biology, Economics, Physics, and Social Science , we will list some, hoping to give student a chance to discover their mathematical formulation:

- Radioactive Decay.
- Population Growth.
- Newton's Law of Cooling.
- Mixture Problems.
- Orthogonal Trajectory.
- Falling Bodies.
- Fluid Flow.
- Simple Electrical Network.
- Linear Rate Equations.
- Compound Interest.

After, a mathematical formulation to the physical problem, as in the case of falling object, which lead to first order differential equation or initial value problem (2.5), we shall learn how to obtain a solution to these equations. We begin by studying separable equations, then exact equations, and then linear equations, and the method of solving these 3-types of equations is the most basic one.

2.2 Separable Equations

These are the simplest type of equations that can be solved using integration.

Definition – 1 : If a linear equation is written in the form,
$$\frac{dy}{dx} = f(x)g(y) \qquad (2.6)$$
Then it can be easily rearranged as,
$$\frac{dy}{g(y)} = f(x)dx. \qquad (2.7)$$

Where the right-hand -side of the equation depends on y only, and the left-hand-side depends on x only, and this is called a separable equation which can be solved by integration.

Definition – 2 : If $F = f(x)$ is a function of x alone, and $G = g(y)$ is a function of y alone, then the differential equation is said to be linear and can be written in the form,
$$f(x)dx + g(y)dy = 0. \qquad (2.8)$$
Which can be solved by integration, and gives implicit solution.

Example – 1 :
Solve the equation by separating the variables:
$$\frac{dy}{dx} = \frac{x}{y}. \qquad (2.9)$$

2.2. SEPARABLE EQUATIONS

Solution:
It is so easy to separate the two functions x, and y from each other,
$$ydy = xdx. \tag{2.10}$$
This separable equation can be solved by integrating both sides with respect of y on the left side, and x on the right side,
$$\int ydy = \int xdx. \tag{2.11}$$
Integrating both sides gives an implicit solution,
$$\frac{y^2}{2} = \frac{x^2}{2} + c. \tag{2.12}$$
To get an explicit solution, means to solve for y as a function of x,
$$y = \pm\sqrt{x^2 + k}. \tag{2.13}$$
Where, $k = 2c$, and c is a constant.

Suppose example-1 is changed as follows:
$$\frac{dy}{dx} = \frac{y}{x}. \tag{2.14}$$
Separating this equation gives,
$$\frac{dy}{y} = \frac{dx}{x}. \tag{2.15}$$
Integrating both sides gives,
$$ln(y) = ln(x) + c \tag{2.16}$$
Then the explicit solution is :
$$y = e^{ln(x)+c} = e^{ln(x)}.e^c = ke^{ln(x)} = kx. \tag{2.17}$$

Where, $k = e^c$, and $e^{ln(x)} = x$.

Now, changing example-1 again as follows:

$$\frac{dy}{dx} = \frac{y}{2-x}. \tag{2.18}$$

Separating this equation gives,

$$\frac{dy}{y} = \frac{dx}{2-x}. \tag{2.19}$$

Integrating both sides gives,

$$ln(y) = -ln(2-x) + c \tag{2.20}$$

Then the explicit solution is :

$$y = e^{-ln(2-x)+c} = e^{-ln(2-x)}.e^c = ke^{-ln(2-x)}. \tag{2.21}$$

Simplifying, the explicit solution is:

$$y(x) = \frac{k}{2-x}.$$

Where, $k = e^c$.

Example – 2 :
Solve the equation by separating the variables:

$$\frac{dy}{dx} = \frac{x}{e^y}. \tag{2.22}$$

Solution:
Separating this equation gives,

$$e^y dy = xdx. \tag{2.23}$$

2.2. SEPARABLE EQUATIONS

Integrating both sides gives,

$$e^y = \frac{x^2}{2} + c \qquad (2.24)$$

Then the explicit solution is :

$$y = ln\{\frac{x^2}{2} + c\}. \qquad (2.25)$$

Example – 3 : IVP
Solve the initial value problem, by separating the variables:

$$\frac{dx}{dt} = \frac{e^{x^2-t}}{x}; x(0) = 0. \qquad (2.26)$$

Solution:
Separating this equation gives,

$$xe^{-x^2} dx = e^{-t} dt. \qquad (2.27)$$

Then the implicit solution is :

$$-\frac{1}{2}e^{-x^2} = -e^{-t} + c. \qquad (2.28)$$

Now, applying the initial conditions $x(0) = 0$, to find the value of c, we get $c - \frac{1}{2}$. The final solution is:

$$-\frac{1}{2}e^{-x^2} = -e^{-t} + \frac{1}{2}. \qquad (2.29)$$

Then the general solution for differential equation (2.26) can be written as,

$$e^{-x^2} = 2e^{-t} - 1. \qquad (2.30)$$

After this practice on solving separable equations, **now we are ready to solve the IVP in (2.5)**, and find the velocity of the falling object:

$$m\frac{dv}{dt} = mg - kv, \ v(0) = v_0. \tag{2.31}$$

For simplification divide (2.31) by m,

$$\frac{dv}{dt} = g - cv. \tag{2.32}$$

Where, $c = \text{constant} = \frac{k}{m}$. then separating equation (2.32) gives,

$$\frac{dv}{g - cv} = dt. \tag{2.33}$$

Integrating both sides finitely by applying the initial conditions: $v(0) = v_0$,

$$\int_{v_0}^{v} \frac{dv}{g - cv} = \int_0^t dt \tag{2.34}$$

$$ln(g - cv) - ln(g - cv_0) = -ct. \tag{2.35}$$
$$ln\frac{g - cv}{g - cv_0} = -ct. \tag{2.36}$$

Then the velocity of the falling object is,

$$v = (v_0 - \frac{g}{c})e^{-ct} + \frac{g}{c}. \tag{2.37}$$

And, is given explicitly as a function of time t.

2.3 Exact Equations

From our discussions so far, we can write the first order differential equation in the following form,

$$\frac{dy}{dx} = -\frac{G(x,y)}{H(x,y)}. \tag{2.38}$$

Separating this equation forms,

$$G(x,y)dx = -H(x,y)dy. \tag{2.39}$$

We may also express this in different form as,

$$G(x,y)dx + H(x,y)dy = 0. \tag{2.40}$$

The differential equation (2.40) is said to be:
Exact Equation in the domain, if and only if there exist a function $F(x,y)$ with total differential,

$$dF(x,y) = G(x,y)dx + H(x,y)dy = 0 \tag{2.41}$$

For all x, and y in the domain. If such a function F exists, then the general solution for the differential equation $dF(x,y) = 0$ can explicitly be given as: $F(x,y) = C$.
To solve the exact equation (2.40), we need to know if the left-hand-side is a total differential. The total differential $dF(x,y)$ of a function $F(x,y)$ of two variables is defined by:

$$\begin{aligned} dF(x,y) &= G(x,y)dx + H(x,y)dy \\ &= \frac{\partial F}{\partial x}(x,y)dx + \frac{\partial F}{\partial y}(x,y)dy. \end{aligned}$$

In solving the exact differential equations, all we need is to find the function F whose total differential dF is

$Gdx + Hdy$. Unfortunately, sometimes it is not simple to determine $F(x, y)$ by inspection, in case like this we need to test the equation for exactness in the region, and this is done simply by showing that :

$$\frac{\partial G}{\partial y}(x, y) = \frac{\partial H}{\partial x}(x, y). \qquad (2.42)$$

For all (x, y) in the region.
To continue with the method, we assume that,

$$\frac{\partial F}{\partial x} = G(x, y), \qquad (2.43)$$

so we can find F by integrating $G(x, y)$ with respect to x, while holding y constant. We write,

$$F(x, y) = \int G(x, y) dx + g(y), \qquad (2.44)$$

where, $g(y)$ is an arbitrary constant. Differentiating (2.44) with respect to y and assume $\partial F/\partial y = H(x, y)$,

$$\frac{\partial F}{\partial y} = \frac{\partial}{\partial y} \int G(x, y) dx + g'(y) = H(x, y). \qquad (2.45)$$

Then,

$$g'(y) = H(x, y) - \frac{\partial}{\partial y} \int G(x, y) dx \qquad (2.46)$$

Integrating equation (2.46) with respect to y and substituting into (2.44). Then the final solution to equation (2.40) is:

$$F(x, y) = C. \qquad (2.47)$$

2.3. EXACT EQUATIONS

Example – 4 : Exact Equation (by Inspection)
Solve the differential equation,

$$(6x\cos y + 1)dx - (3x^2 \sin y + 1)dy = 0. \qquad (2.48)$$

Solution:
Rearranging (2.48) as follows,

$$(6x\cos y\, dx - 3x^2 \sin y\, dy) + dx - dy = 0. \qquad (2.49)$$
$$\text{Gives,}\ d(3x^2 \cos y) + dx - dy = 0. \qquad (2.50)$$

Which is,

$$d(3x^2 \cos y + x - y) = 0. \qquad (2.51)$$

then the function is:

$$F(x, y) = 3x^2 \cos y + x - y, which\ satisfies, \qquad (2.52)$$

$$\frac{\partial F}{\partial x} = 6x\cos y + 1.$$

$$\frac{\partial F}{\partial y} = -3x^2 \sin y - 1.$$

Then equation (2.48) is exact equation, with solution,

$$3x^2 \cos y + x - y - c. \qquad (2.53)$$

To check the exactness, we have to show that :
$\partial G/\partial y = \partial H/\partial x$, first.

$$G(x, y) = (6x\cos y + 1),\ \text{and}\ \frac{\partial G}{\partial y} = -6x \sin y.$$

$$H(x, y) = -(3x^2 \sin y - 1),\ \text{and}\ \frac{\partial H}{\partial x} = -6x \sin y.$$

Then the exactness is proved.

Example – 5 : Exact Equation
Solve the differential equation,

$$(sinx + 2xy)dx + (x^2 + 2y)dy = 0. \qquad (2.54)$$

Solution:
First we will find the functions $G(x,y)$, $H(x,y)$, and their differentials: $\frac{\partial G}{\partial y}$, and $\frac{\partial H}{\partial x}$, then test their exactness,

$$G(x,y) = (sinx + 2xy), \text{ and } \frac{\partial G}{\partial y}(x,y) = 2x.$$

$$H(x,y) = (x^2 + 2y), \text{ and } \frac{\partial H}{\partial x}(x,y) = 2x.$$

Then, equation (2.54) is an exact equation. The next step is to find the total function $F(x,y)$ by integration as follows,

$$F(x,y) = \int G(x,y)dx + f(y). \qquad (2.55)$$

Or,

$$F(x,y) = \int H(x,y)dy + f(x). \qquad (2.56)$$

We need only to compute one of them. Substituting we get:

$$F(x,y) = \int (sinx + 2xy)dx + f(y). \qquad (2.57)$$
$$= -cosx + x^2 y + f(y) \qquad (2.58)$$

Taking the partial derivative with respect to y and substitute $(x^2 + 2y)$ for H:

$$\frac{\partial F}{\partial y}(x,y) = H(x,y). \qquad (2.59)$$
$$x^2 + f'(y) = x^2 + 2y. \qquad (2.60)$$

2.3. EXACT EQUATIONS

Thus, (2.60) gives, $f'(y) = 2y \Rightarrow f(y) = y^2$.

Substituting back we get :
$$F(x,y) = -\cos x + x^2 y + y^2. \qquad (2.61)$$
Then the solution to equation(2.54) is:
$$x^2 y - \cos x + y^2 = C. \qquad (2.62)$$
Which is the implicit solution.

Note: Sometimes an equation that is not exact equation can be changed to exact equation with simple manipulation. For example the equation:
$(x^2 + 3x^4 \sin y)dx + (x^5 \cos y)dy = 0$ is not exact equation, but multiplying it by the factor x^{-2} will change it into Exact equation. This is left for the student to solve.

Example – 6 : Exact Equation
Determine whether the given equation is an exact equation, if it is, then solve it:
$$\sin\theta\, dr + (r\cos\theta - e^\theta)\, d\theta = 0. \qquad (2.63)$$

Solution:

$$G(r,\theta) = (\sin\theta), and\ \frac{\partial G}{\partial \theta}(r,\theta) = \cos\theta.$$

$$H(r,\theta) = (r\cos\theta - e^\theta), and\ \frac{\partial H}{\partial r}(r,0) = \cos\theta.$$

Since $\frac{\partial G}{\partial \theta}(x,y) = \frac{\partial H}{\partial r}(x,y)$, This proves that equation (2.63) is an exact equation. Now we need to obtain the function $F(r,\theta)$,

$$\begin{aligned} F(r,\theta) &= \int (\sin\theta) dr + f(\theta). \\ &= r\sin\theta + f(\theta). \end{aligned}$$

Taking the partial derivative with respect to θ and substitute $-(r\cos\theta - e^\theta)$ for H as follows,

$$\frac{\partial F}{\partial \theta}(r,\theta) = H(r,\theta).$$
$$r\cos\theta + f'(\theta) = r\cos\theta - e^\theta.$$

then, $f'(\theta) = e^\theta$, and $\Rightarrow f(\theta) = -e^\theta$. Substituting back gives,

$$F(r,\theta) = r\sin\theta - e^\theta, \tag{2.64}$$

and letting it equal a constant C, then the implicit solution to equation (2.63) is,

$$r\sin\theta - e^\theta = C. \tag{2.65}$$

Or $\Rightarrow r = (C + e^\theta)\csc(\theta)$.

2.4 Linear Equations

The Integrating Factor

As we have mentioned before, the first order ordinary differential equations are divided into two classes:

- Linear ordinary differential equations, and
- Non-Linear ordinary differential equations.

The linear first order equation is an equation that can be expressed in a standard form as,

$$\frac{dy}{dx} + P(x)y = Q(x). \tag{2.66}$$

Where, $P(x)$, and $Q(x)$ are continuous functions in the region. Testing this equation for exactness, proves to be

2.4. LINEAR EQUATIONS

exact when $P(x) \equiv 0$, and its solution leads to,

$$I(x) = e^{\int P(x)dx}. \tag{2.67}$$

Which is called the integrating factor, and this plays an essential part in solving the linear equations. Solving for the dependent variable y equation (2.66) yields:

$$y(x) = \frac{\int I(x)Q(x)\,dx + c}{I(x)}. \tag{2.68}$$

This function is known as the **general solution** to the linear equation (2.66).

To describe the method of solution of linear equations by using the integrating factor would be easier to solve a regular problem in steps.

Example – 7 : Linear Equations
Find the general solution to the differential equation:

$$\frac{dy}{dx} = x^2 y + 3x^2. \tag{2.69}$$

Step-1: Rewrite the equation in the standard form as in (2.66):

$$\frac{dy}{dx} - x^2 y = 3x^2. \tag{2.70}$$

From this equation $P(x) = -x^2$, and $Q(x) = 3x^2$.
Step-2:
Calculate the integrating factor $I(x)$:

$$\begin{aligned} I(x) &= e^{\int p(x)dx} \\ &= e^{-\int x^2 dx} \\ &= e^{-x^3/3}. \end{aligned}$$

Step-3::
Multiplying this integrating factor $I(x)$ by the equation in standard form (2.70):

$$\underbrace{(e^{-x^3/3}).\frac{dy}{dx} - (e^{-x^3/3}).x^2 y}_{\frac{d}{dx}(I(x)y)} = (e^{-x^3/3}).3x^2. \qquad (2.71)$$

Where the left-side of the equation is just $\frac{d}{dx}(I(x)y)$ or differentiation with respect to the dependent variable of (the Integrating factor × the dependent variable). Then equation (2.71) can be written as,

$$\frac{d}{dx}(e^{-x^3/3})y = (e^{-x^3/3}).3x^2. \qquad (2.72)$$

Step-4::
Integrate both sides of the equation,

$$\begin{aligned}(e^{-x^3/3})\, y &= \int (e^{-x^3/3}).3x^2.dx. \\ &= -3e^{-x^3/3} + C.\end{aligned}$$

Step-5::
Solve for y,

$$\begin{aligned} y &= 3(e^{-x^3/3})(e^{x^3/3}) + C(e^{x^3/3}). \\ &= 3 + C(e^{x^3/3}). \end{aligned}$$

Thus the solution for the differential equation (2.69) is,

$$y(x) = 3 + C(e^{x^3/3}).$$

And this solution is valid for all x, since both $P(x) = -x^2$, and $Q(x) = 3x^2$ are continuous for all x.

2.4. LINEAR EQUATIONS

Example – 8 : IVP

Solve the initial value problem,

$$x^4 \frac{dy}{dx} + 4x^3 y = x^2; \ y(3) = 0. \qquad (2.73)$$

Solution:

To solve this IVP we will follow all the steps as done in example-1 from step-1 to step-5, and with extension of one more step, step-6 in which we will compute the value of the unknown constant C by using the given initial conditions.

Start with rewriting the equation in the standard form:

$$\frac{dy}{dx} + \frac{4}{x} y = \frac{1}{x^2}. \qquad (2.74)$$

From this equation we get $P(x) = \frac{4}{x}$, and $Q(x) = \frac{1}{x^2}$, then we can calculate the integrating factor,

$$\begin{aligned} I(x) &= e^{\int P(x)dx} = e^{\int \frac{4}{x} dx}. & (2.75) \\ &= e^{4 \ln(x)} = e^{\ln(x^4)} = x^4. & (2.76) \end{aligned}$$

Multiplying the integrating factor $I(x) = x^4$ by the equation in the standard form (2.74),

$$\underbrace{(x^4).\frac{dy}{dx} + (x^4).\frac{4}{x} y}_{\frac{d}{dx}(x^4 y)} = (x^4).\frac{1}{x^2}. \qquad (2.77)$$

As you can see the left-side of the equation is just the derivative (with respect to x) of the integrating factor times the dependent variable y.

$$\begin{aligned} \frac{d}{dx}(x^4 y) &= (x^4).\frac{1}{x^2}. \\ &= x^2. \end{aligned}$$

Integrating this equation gives,

$$x^4 y = \frac{x^3}{3} + C. \tag{2.78}$$

Then solving for y gives,

$$y = \frac{x^{-1}}{3} + cx^{-4}. \tag{2.79}$$

This is the general solution in terms of unknown constant C. To find the value of C we use the given initial conditions $y(3) = 0$ to get the final solution:

$$y(3) = 0 = \frac{3^{-1}}{3} + C(3^{-4}), where\ C = -9. \tag{2.80}$$

Then, the solution to the IVP in (2.73) is,

$$y(x) = \frac{x^{-1}}{3} - 9x^{-4}. \tag{2.81}$$

2.5 Homogeneous Equations

Solving differential equations of the form,

$$G(x,y)dx + H(x,y)dy = o. \tag{2.82}$$

We were able to solve it as a separable, exact, or linear equation. Unfortunately, there are cases where the variables of the differential equation of this form are not separable, exact, or linear equations, but it may still be possible to transform it into a separable, exact, or linear by means of a suitable substitution or transformation. This class of equations is called **homogeneous equations**.

2.5. HOMOGENEOUS EQUATIONS

Definition – 1 : Homogeneous Equations
If the differential equation (2.82) has a property that:

$$G(\delta x, \delta y) = \delta^n G(x, y).$$
$$\text{And, } H(\delta x, \delta y) = \delta^n H(x, y).$$

Then the differential equation is a homogeneous equation.

Definition – 2 : Homogeneous Equations
For the differential equation,

$$\frac{dy}{dx} = f(x, y). \qquad (2.83)$$

If the right-hand-side of the equation can be expressed as a function of the fraction (y/x), then the equation is called a Homogeneous Equation.

In what follow we will present some problems, and show the way to transform them into a suitable and solvable equations.

Exapmple – 9:
Transform the Non-homogeneous equation into a homogeneous equation:

$$(x + y)dx + xdy = 0. \qquad (2.84)$$

Solution:
Rewriting this equation as,

$$\frac{dy}{dx} = -\frac{x+y}{x} = -(1 + \frac{y}{x}).$$
$$\frac{dy}{dx} + (\frac{1}{x})y = -1.$$

Where the last form is a homogeneous equation.

Exapmple – 10:
Transform the differential equation into a homogeneous equation:
$$(x+y-3)dx - (x+y)dy = 0. \qquad (2.85)$$

Solution:
rewriting and dividing by x,
$$\frac{dy}{dx} = \frac{x+y-3}{x+y} = \frac{1+y/x-3/x}{1+y/x}.$$
The last form is still non-homogeneous.

Testing for Homogeneity:
Thus, dealing with differential equations, we have to test for homogeneity, and is done as follows: on the given equation of the form,
$$\frac{dy}{dx} = f(x,y). \qquad (2.86)$$
Replace x with δx, and y with δy, then if: $f(x,y) = \delta^n f(\delta x, \delta y)$, then the function $f(x,y)$ is said to be homogeneous of degree n.

Transforming Non – Homogeneous into Homogeneous
If the differential equation :
$$\frac{dy}{dx} = f(x,y). \qquad (2.87)$$
Proves to be non-homogeneous, then a substitution of the following types,
$$y = ux, \text{ with } \frac{dy}{dx} = u + x\frac{du}{dx}. \text{ Or} \qquad (2.88)$$

2.5. HOMOGENEOUS EQUATIONS

$$x = vy, \text{ with } \frac{dx}{dy} = v + y\frac{dv}{dy}. \qquad (2.89)$$

Will transform it into a Homogeneous one in which the variables are separable. This is illustrated in the following examples:

Exapmple – 11:
The differential equation:

$$\frac{dx}{dt} = \frac{t^2 + x^2 + x}{t}, \qquad (2.90)$$

is non-homogeneous equation. But substituting the following transformation,

$$x = ut, \quad \frac{dx}{dt} = u + t\frac{du}{dt}, \qquad (2.91)$$

into the differential equation (2.90), and simplifying gives,

$$u + t\frac{du}{dt} = \frac{1}{t}(t^2 + u^2 t^2 + ut).$$
$$\frac{du}{dt} = (1 + u^2).$$

a separable equation, which can be solved by separation as follows:

$$\frac{du}{1 + u^2} = dt.$$
$$\int \frac{du}{1 + u^2} = \int dt.$$
$$tan^{-1} u = t + c.$$
$$u = tan(t + c).$$
$$\frac{x}{t} = tan(t + c).$$
$$\text{Then, } x(t) = t\, tan(t + c).$$

2.6 Non-Linear equations and Solutions

As mentioned before, the Linear Differential equation of first order in general has the following form:

$$\frac{dy}{dx} + P(x)(x)\frac{dy}{dx} = Q(x). \tag{2.92}$$

And any equations that does not match this equation is considered non-linear differential equations.

Remarks: For non-linear differential equations of first order but with degree $n > 1$ s.a $(\frac{dy}{dx})^2$, three types of solution can be applied:
1. Equations solvable in x.
2. Equations solvable in y.
3. Equations solvable in y' or $(\frac{dy}{dx})$.

1. **Non-Linear DE Solvable in x**

Example – 12 : Solve the following non-linear differential equation:

$$(\frac{dy}{dx})^2 + 2x(\frac{dy}{dx}) - y = 0. \tag{2.93}$$

Solution:
To solve this equation, the following steps are used:
1. replace $\frac{dy}{dx}$ with q.

$$q^2 + 2xq - y = 0.$$

2. Solve for x.

$$x = \frac{y - q^2}{2q}.$$

2.6. NON-LINEAR EQUATIONS AND SOLUTIONS

3. Differentiate with respect to y.

$$\frac{dx}{dy} = \frac{2q(1 - 2q\frac{dq}{dy}) - (y - q^2).2\frac{dq}{dy}}{4q^2}.$$

4. replace $\frac{dx}{dy}$ with $\frac{1}{q}$, and simplify.

$$2q = q - q^2\frac{dq}{dy} - y\frac{dq}{dy}.$$
$$2qdy = qdy - (q^2 + y)dq.$$
$$(q^2 + y)dq + qdy = 0.$$

By inspection the last equation is:

$$d(\frac{q^3}{3} + yq) = 0.$$

And with differentiation gives,

$$\frac{q^3}{3} + yq = c.$$

Then the general solution is:

$$y(x) = \frac{k - q^3}{3q}, \quad where \; k = 3c. \tag{2.94}$$

2. Non-Linear DE Solvable in y

Example - 13 : Using the same non-linear differential equation:

$$(\frac{dy}{dx})^2 + 2x(\frac{dy}{dx}) - y = 0. \tag{2.95}$$

And following similar steps:
1. replace $\frac{dy}{dx}$ with q.

$$q^2 + 2xq - y = 0.$$

2. Solve for y.
$$y = q^2 + 2xq.$$

3. Differentiate with respect to x.
$$\frac{dy}{dx} = 2q\frac{dq}{dx} + 2q + 2x\frac{dq}{dx}.$$

4. replace $\frac{dx}{dy}$ with q, and simplify.
$$q = 2qx\frac{dq}{dy} + 2q + 2x\frac{dq}{dx}.$$
$$2xdq + 2qdq + qdx = 0.$$

By inspection the last equation is:
$$d(q^2x + 2/3q^3) = 0.$$

And with differentiation gives,
$$q^2x + 2/3q^3 - c.$$

Then the general solution in x is:
$$x = \frac{k - 2q^3}{3q^2}, \quad \text{where } k = 3c. \tag{2.96}$$

3. **Non-Linear DE Solvable in y'**

Example-14: Solve the following non-linear differential equation:
$$2y(\frac{dy}{dx})^2 + (6x^2y - 1)(\frac{dy}{dx}) - 3x^2 = 0. \tag{2.97}$$

2.6. NON-LINEAR EQUATIONS AND SOLUTIONS

To solve this equation, the following steps are used:
1. replace $\frac{dy}{dx}$ with q.

$$2yq^2 + (6x^2y - 1)q - 3x^2 = 0. \tag{2.98}$$

Notice equation (2.98) is a quadratic equation in q. Solving this equation for q using quadratic formula gives,

$$q = \frac{(1 - 6x^2y) \pm (6x^2y + 1)}{4y}$$

Then the roots of the equation are:

$$q_1 = \frac{1}{2y}.$$
$$q_2 = -3x^2.$$

But $q = \frac{dy}{dx}$. Then the two solutions are:

$$y^2 - x + c = 0.$$
$$y + x^3 + c = 0.$$

And the general solution is:

$$(y + x^3 + c)(y^2 - x + c) = 0.$$

Bernoulli's Equation

Historical Notes

James Bernoulli (1654 − 1705), a swiss mathematician, studied Calculus of Newton and Leibniz, was a professor of Mathematics at Basel University. In 1695 he proposed the equation: $y' + p(x)y = qy^n$, the equation was solved by his brother John Bernoulli(1667 − 1748) who became a Mathematician, and applied Calculus to Geometry and differential equations. In 1696 Leibniz showed that Bernoulli's equation can be reduced to a linear equation by making the substitution of : $u = y^{1-n}$.

Here we start with "Bernoulli's equation", as a special type of first order differential equations with non-linear form,

$$\frac{dy}{dx} + P(x)y = Q(x)y^m, \qquad (2.99)$$

where, m is any real number. Notice, that equation (2.99) can be linear equation if $m = 0$ or 1.
To transform this equation into a linear one, first we try to get rid of y^m on the right-side, by multiplying the entire equation by y^{-m}.

$$y^{-m}\frac{dy}{dx} + P(x)y^{1-m} = Q(x) \qquad (2.100)$$

Then we will chose some variable $u = y^{1-m}$, and differentiate u with respect to x (independent variable),

$$\begin{aligned} u &= y^{1-m}. \\ \frac{du}{dx} &= (1-m)y^{-m}\frac{dy}{dx}. \end{aligned}$$

2.6. NON-LINEAR EQUATIONS AND SOLUTIONS

Substituting back into equation (2.100) gives,

$$(1-m)^{-1}\frac{du}{dx} + P(x)u = Q(x).$$
$$\frac{du}{dx} + \frac{P(x)}{1-m}u = Q(x).$$

Which is a linear equation.

Example of Bernoulli equation type:
Solve the following Bernoulli Equation,

$$\frac{dy}{dx} + 4y = xy^{-4}. \tag{2.101}$$

First, we will multiply the equation by y^4,

$$y^4\frac{dy}{dx} + 4y^5 = x. \tag{2.102}$$

Then chose $u = y^5$, and differentiate with respect to x.

$$u = y^5.$$
$$\frac{du}{dx} = 5y^4\frac{dy}{dx}.$$

Then substituting back into equation (2.102) gives,

$$\frac{du}{dx} + 20u = 5x.$$

Which is a linear equation with a solution.

Note: The solution of this equation requires finding the integrating factor $I(x)$ first. The solution is left for the student as a practice.

2.7 Applications of First Order Equations

Mixture Problems

Th basic mixture problem with one compartment system consists of a function $A(t)$ which represents the amount of the solution in the compartment at time t, another solution of some concentration (c_i) enters the compartment with an input rate (r_i), the two solutions then are well mixed, and flows out with concentration (c_o) and rate (r_o). Since the rate of change in the amount of the substance ($\frac{dA(t)}{dt}$) in time t depends on the difference between the input rate and output rate, then the system suggests:

$$\frac{dA}{dt} = \text{input rate} - \text{output rate} \qquad (2.103)$$

We will be using this as a mathematical model for the process. Since input rate is $c_i r_i$, and the output rate is $c_o r_o$, then the above mathematical model for mixture problems can be written as:

$$\frac{dA}{dt} = c_i r_i - c_o r_o. \qquad (2.104)$$

Problem – 1 : When $r_i = r_o$
A tank containing $500L$ of water. A salt solution of amount $s(t)$, and concentration $(.5kg/L)$ Flows into the tank with input rate of $(3L/min)$, after being well stirred inside the tank it flows out of the tank with output rate equal to the input rate. Determine the output concentration of the mixture.

```
r_i = 3L/min  ─────▶         s(t)
c_i = .5kg/L          ┌──────────────┐
                      │    500L      │
                      │  s(0) = 0    │──── r_o = 3L/min
                      └──────────────┘     c_o = ?
```

$r_i = 3L/min$
$c_i = .5kg/L$
$s(t)$
$500L$
$s(0) = 0$
$r_o = 3L/min$
$c_o = ?$

Fig(1)- Mixing Problem with equal flow rates

Solution :

From the given information we can calculate the input flow $r_i c_i$:

$$r_i c_i \;=\; (3L/min\,>)(5kg/L) = 1.5kg/min.$$

Since the in-put rate r_i and the out-put rate r_o are the same ($=3L/min$), then there will be no change in the concentration of the out-flow which is:

$$c_o = \frac{amount\ of\ the\ solution}{volume\ of\ the\ solution} = \frac{s(t)}{500}.$$

Then the out-put flow $r_o c_o$ is:

$$r_o c_o \;=\; 3(\frac{s(t)}{500}).$$

Then, using the above model, or equation (2.104) we get.

$$\frac{ds}{dt} = 1.5 - \frac{3s}{500}. \qquad (2.105)$$

Equation (2.105) is a linear and separable equation which can be solved easily using the method studied before, and with the help of the initial conditions $s(0) = 0$ we can evaluate the arbitrary constant in the problem. Thus the IVP here is:

$$\frac{ds}{dt} = 1.5 - \frac{3s}{500}; \text{ with } s(0) = 0 \qquad (2.106)$$

Simplifying equation (2.106) gives,

$$\frac{ds}{dt} = \frac{750 - 3s}{500} = 3\frac{250 - s}{500}. \qquad (2.107)$$

Then the separable equation is,

$$\frac{ds}{250 - s} = \frac{3}{500} dt. \qquad (2.108)$$

Integrating both sides,

$$\int \frac{ds}{250 - s} = \int \frac{3}{500} dt.$$
$$ln(250 - s) = -\frac{3t}{500} + c.$$
$$250 - s = e^{-3t/500 + c}$$
$$= ke^{-3t/500}, \text{ where } k = e^c.$$
$$\text{Then, } s(t) = 250 - ke^{-3t/500}.$$

Applying the initial conditions, $s(0) = 0$ to find k we get $k = 250$, then the amount of the solution at any time is:

$$s(t) = 250 - 250e^{-3t/500} \qquad (2.109)$$

Since the output concentration is $c_o = (s(t)/500)$.

$$\text{Then, } c_o = 1/2(1 - e^{-3t/500}) kg/L. \qquad (2.110)$$

Problem – 2 : When $r_i > r_o$

Suppose we use the same problem-1 but with the difference in the inflow rate $= 4L/min$ which is larger than the outflow rate. We would like to determine the output concentration of the mixture as a function of time.

$r_i = 4L/min$
$c_i = .5kg/L$

$s(t)$
L?
$s(0) = 0$

$r_o = 3L/min$
$c_o = ?$

Fig(2)- Mixing Problem with different flow rates

- - - -

Solution :

In this problem the in-put rate r_i is larger than the output r_o rate, and the difference is:

$$r_i - r_o = 4L/min - 3L/min = 1L/min.$$

And this difference in rate will change the volume of the mixture inside the tank after t minutes into:

$$V(t) = initial volume + change in volume.$$
$$= 500 + 1t.$$

This in turn will change the concentration of the out-flow as follows:

$$c_o = \frac{amount\ of\ the\ solution}{volume\ of\ the\ solution} = \frac{s(t)}{500 + 1t}.$$

Hence, the out-flow rate becomes:

$$r_o c_o = 3\left(\frac{s(t)}{(500+t)}\right).$$

And the in-flow rate is:

$$r_i c_i = (4L/min)(.5kg/L) = 2kg/min.$$

Then the model for the system is:

$$\frac{ds(t)}{dt} = input\ rate - output\ rate. \qquad (2.111)$$

$$\frac{ds(t)}{dt} = 2 - \frac{3s(t)}{(500+t)}. \qquad (2.112)$$

Equation (2.112) is not a separable equation, but it is a linear differential equation, then we can use the method of integrating factor that was used in solving the linear equations in $S(2.3)$, also the problem in here is initial value problem with initial conditions $s(0) = 0$,

$$\frac{ds(t)}{dt} = 2 - \frac{3s(t)}{(500+t)}; with\ s(0) = 0. \qquad (2.113)$$

To find the integrating factor, first we have to rewrite the equation into a linear form as,

$$\frac{ds(t)}{dt} + \frac{3s(t)}{(500+t)} = 2. \qquad (2.114)$$

Where, $p(t) = \frac{3}{(500+t)}$, and the integrating factor is,

$$I(t) = e^{\int P(t)dt} = e^{\int \frac{3}{(500+t)}dt} = (500+t)^3. \qquad (2.115)$$

2.8. TECHNICAL WRITING

Multiplying the integrating factor $I(t)$ by each term of equation (2.114) and simplifying gives,

$$\frac{d}{dt}((500+t)^3 s) = 2(500+t)^3 dt.$$

$$(500+t)^3 s = \frac{(500+t)^4}{2} + C.$$

$$s(t) = \frac{(500+t)^4}{2} + C(500+t)^{-3}.$$

Applying the initial conditions $s(0) = 0$ gives,

$$C = \frac{-(500^4)}{2}.$$

Then the amount of solution in the tank at anytime t will be,

$$s(t) = \frac{(500+t)}{2} - \frac{(500)^4}{2}(500+t)^{-3}. \qquad (2.116)$$

2.8 Technical Writing

Student will refer to page (24) from chapter-1, chose a topic from the list, Write a problem, and formulate a model for the problem, and solve the problem, or IVP if possible. A good witting will be done in detail description, with schematic diagrams if any.

2.9 Review Exercises

Separable Equations: In problems 1 – 6 separate the differential equations, and solve:

1. $\dfrac{dy}{dx} = 2(x-4).$

2. $\dfrac{d\theta}{dt} = \dfrac{(1+t)}{\theta^2}.$

3. $\dfrac{dx}{dt} = e^{-x} \sin t.$

4. $\dfrac{dy}{dx} = \sqrt{2 - x + 2y - xy}.$

5. $\dfrac{dy}{dx} = \dfrac{y \log y}{3x}.$

6. $\dfrac{dy}{dx} = \dfrac{y e^x}{1 - e^x}.$

7. Solve the initial value problem:
$$\dfrac{dy}{dt} = 2(x-4); \; y(0) = 2.$$

8. Solve the IVP :
$$\dfrac{dy}{dt} = \dfrac{t^3}{y^2}; \; y(0) = 0.$$

2.9. REVIEW EXERCISES

In problems 9–13 determine whether the given equations are exact, if exact solve it:

9. $(2x+4)dx - (3t-1)dt = 0.$

10. $2x + (2t - 5x)x' = 0.$

11. $(2xt^2 + 4)dx - (3 - 2x^2 t)dt = 0.$

12. $2xy\,dx - x^2 dy = 0.$

13. $y' = 2e^x - \dfrac{y}{x} + 6x.$

Linear Equations :
Find the general solution for the problems 14 – 17:

14. $4y' + 2y = 0.$

15. $y' + 3y/x = 5.$

16. $y' + 2y = 0.$

Bernoulli's Equations :
Solve the following Bernoulli type equations:

17. $y' + y = xy^4.$

18. $y' + y^2/x^2 = y/x.$

19. $xy^2 y' + 2y^3 = 1.$

20. Solve the initial value problem:

$$\frac{dy}{dx} + \frac{x}{2y^2} = \frac{y}{x}; \quad y(1) = 2.$$

Mixture Problems :
Solve the following mixture problems:

21. A large tank holding $400gal$ shown in ($fig.4$) of water which contains initially 60pounds of salt dissolved in it. A brine solution of concentration of $2lb/gal$ is pumped inside the tank at a rate of $4gal/min$, then after being well stirred it is pumped out of the tank with the same rate. Determine the amount of the salt in the tank at any time.

$r_i = 4gal/min$
$c_i = 2lb/gal$

$s(t)$
$400gal$
$s(0) = 60lb$
$r_o = 4gal/min$
$c_o = ?$

$Fig(3)$ − $Mixing\ Problem\ with\ equal\ flow\ rates$

. . . .

2.10 Review Exercise Solutions

1. $$\frac{dy}{dx} = 2(x-4).$$

Separating the equation and solving gives,

$$dy = 2(x-4)dx.$$
$$\text{Then, } y(x) = x^2 - 8x + c.$$

2.10. REVIEW EXERCISE SOLUTIONS

2. $\dfrac{d\theta}{dt} = \dfrac{(1+t)}{\theta^2}.$

Separating the equation and solving gives,

$$\theta^2 d\theta = (1+t)dt.$$
$$\dfrac{(\theta)^3}{3} = t + \dfrac{t^2}{2} + c$$

3. $\dfrac{dx}{dt} = e^{-x}\sin t.$

Separating the equation and solving gives,

$$e^x dx = \sin t\, dt.$$
$$e^x = -\cos t + c.$$
$$x(t) = k e^{-\cos t}.$$

4. $\dfrac{dy}{dx} = \sqrt{2 - x + 2y - xy}.$
$$= \sqrt{(2-x) + y(2-x)}.$$
$$= \sqrt{(2-x)}\sqrt{(1+y)}$$

Separating the last expression and solving gives,

$$\dfrac{dy}{\sqrt{1+y}} = \sqrt{2-x}\,dx.$$
$$\int \dfrac{dy}{\sqrt{1+y}} = \int \sqrt{2-x}\,dx.$$
$$2\sqrt{1+y} = -\dfrac{2}{3}\sqrt{(2-x)^3} + c.$$
$$2\sqrt{1+y} + \dfrac{2}{3}\sqrt{(2-x)^3} = c.$$

5. $$\frac{dy}{dx} = \frac{y \, logy}{3x}.$$

Separating the last expression and solving gives,

$$\frac{dy}{ylogy} = \frac{dx}{3x}.$$

$$\int \frac{1}{logy} \frac{dy}{y} = \int \frac{1}{3} \frac{dx}{x}.$$

Let $u = logy$, then $du = \frac{dy}{y}$. Substituting back gives,

$$\int \frac{du}{u} = \frac{1}{3} \int \frac{dx}{x}.$$

$$lnu = \frac{1}{3}lnx + c.$$

$$3lnu = ln(cx).$$

Then, $y(x) = e^{(cx)^{1/3}}$.

6. $$\frac{dy}{dx} = \frac{ye^x}{1 - e^x}.$$

Separating the last expression and solving gives,

$$\frac{dy}{y} = \frac{e^x}{1 - e^x} dx.$$

Let $u = 1 - e^x$, then $du = -e^x dx$, substituting back,

$$\int \frac{dy}{y} = -\int \frac{du}{u}.$$

$$lny = -lnu + c.$$

Or, $y(x) = \frac{k}{u}$.

$$y(x) = \frac{k}{1 - e^x}.$$

2.10. REVIEW EXERCISE SOLUTIONS

7. Solve the initial value problem:

$$\frac{dy}{dt} = 2(x-4); \; y(0) = 2.$$

From problem (1) the solution for this problem is:

$$y(x) = x^2 - 6x + c.$$
$$and \; from \; IV \; we \; get, c = 2.$$
$$Then \; the \; final \; solution \; is : y(x) = x^2 - 6x + 2.$$

8. Solve the IVP :

$$\frac{dy}{dt} = \frac{t^3}{y^2}; \; y(0) = 0.$$

Separating and solving gives,

$$y^2 dy = t^3 dt.$$
$$\int y^2 dy = \int t^3 dt.$$
$$y^3/3 = t^4/4 + c.$$

9. $(2x+4)dx + (3t-1)dt = 0.$

Solving this exact equation:

$$G(x,t) = 2x + 4 \Rightarrow \frac{\partial G}{\partial t} = 0.$$
$$H(x,t) = 3t - 1 \Rightarrow \frac{\partial H}{\partial x} = 0.$$

This proves the exactness of the equation. Then,

$$F(x,t) = \int G(x,t)dx + f(t).$$

$$\begin{aligned}
&= \int (2x+4)dx + f(t). \\
F(x,t) &= x^2 + 4x + f(t). \\
\frac{\partial F}{\partial t} &= 0 + f'(t) = 3t - 1. \\
f'(t) &= 3t - 1. \\
f(t) &= \frac{3t^2}{2} - t. \\
\text{Then, } F(x,t) &= x^2 + 4x + \frac{3t^2}{2} - t \\
&= c.
\end{aligned}$$

$$Or, \ x^2 + 4x + \frac{3t^2}{2} - t \ = \ c \ is \ the \ solution.$$

10. $\quad 2x + (2t - 5x)x' = 0.$

$$2x + (2t - 5x)\frac{dx}{dt} = 0.$$
$$2xdt + (2t - 5x)dx = 0.$$

Proving the exactness we find,

$$G(t,x) = 2x \Rightarrow \frac{\partial G}{\partial x} = 2.$$
$$H(t,x) = 2t - 5x \Rightarrow \frac{\partial H}{\partial t} = 2.$$

The equation is an exact equation.

$$\begin{aligned}
F(t,x) &= \int G(t,x)dt + f(x). \\
&= \int 2xdt + f(x) \\
F(t,x) &= 2xt + f(x). \\
\frac{\partial F}{\partial x} &= 2t + f'(x) = 2t - 5x. \\
f'(x) &= -5x \Rightarrow f(x) = \frac{-5x^2}{2}.
\end{aligned}$$

2.10. REVIEW EXERCISE SOLUTIONS

$$\text{Then, } F(t,x) = 2xt - \frac{5x^2}{2}.$$

$$Or, \ 2xt - \frac{5x^2}{2} = c.$$

11. $(2xt^2 + 4)dx - (3 - 2x^2t)dt = 0.$

$$G(x,t) = 2xt^2 = 4 \Rightarrow \frac{\partial G}{\partial t} = 4xt.$$

$$H(x,t) = -3 + 2x^2t \Rightarrow \frac{\partial H}{\partial x} = 4xt.$$

The equation is an exact equation.

$$F(x,t) = \int G(x,t)dx + f(t).$$

$$= \int (2xt^2 + 4)dx + f(t)$$

$$F(x,t) = t^2x^2 + 4x + f(t).$$

$$\frac{\partial F}{\partial t} = 2tx^2 + f'(t)$$

$$= -3 + 2x^2t.$$

$$f'(t) = -3 \Rightarrow f(t) = -3t.$$

$$\text{Then, } F(x,t) = x^2t^2 + 4x - 3t.$$

$$Or, \ t^2x^2 + 4x - 3t = c.$$

12. $2xydx - x^2dy = 0.$

$$G(x,y) = 2xy \Rightarrow \frac{\partial G}{\partial y} = 2x.$$

$$H(x,y) = -x^2 \Rightarrow \frac{\partial H}{\partial x} = -2x.$$

Then the equation is not an exact equation.

13. $\quad y' = 2e^x - \dfrac{y}{x} + 6x.$

Rewriting the equation as:

$$xdy + (y - 2xe^x - 6x^2)dx = 0.$$

$$\begin{aligned} G(y,x) &= x \Rightarrow \dfrac{\partial G}{\partial x} = 1. \\ H(y,x) &= (y - 2xe^x - 6x^2) \\ &\Rightarrow \dfrac{\partial H}{\partial y} = 1. \end{aligned}$$

The equation is an exact equation.

$$F(y,x) = \int G(y,x)dy + f(x).$$
$$= \int xy + f(x)$$
$$\dfrac{\partial F}{\partial x} = y + f'(x) = y - 2xe^x - 6x^2.$$
$$f'(x) = -2xe^x - 6x^2 \Rightarrow f(x)$$
$$= -2xe^x + 2e^x + 4x^3.$$
$$\text{Then, } F(y,x) = xy - 2xe^x + 2e^x + 4x^3.$$
$$\text{Or,} \quad xy - 2xe^x + 2e^x + 4x^3 = c.$$

Linear Equations :

14. $\quad 4y' + 2y = 0.$

Rewrite the equation in the standard form of linear equation:

$$y' + 1/2y = 0.$$

2.10. REVIEW EXERCISE SOLUTIONS

This is a separable equation.
$$dy/y = -dx.$$
$$\text{Then, } y(x) = ke^{-x/2}.$$

15. $y' + 3y/x = 5.$

from the linear equation we get $P(x) = 3/x$, and the integrating factor $I(x) = e^{\int P(x)dx} = x^3$. Multiplying this integrating factor by equation (15) and simplifying to get the solution:

$$x^3 y' + x^3 3y/x = 5x^3.$$
$$\underbrace{x^3 y' + x^3 3y/x}_{\frac{\partial}{\partial x} x^3 y} = 5x^3.$$

$$\text{Integrating both sides gives}: x^3 y = \frac{5}{4}x^4 = c.$$

$$\text{Then, the solution is}: y(x) = \frac{5}{4}x + cx^{-3}.$$

16. $y' + 2y = 0.$

In this equation the $p(x) = 2$, and the integrating factor is: $I(x) = e^{\int p(x)dx} = e^{\int 2dx} = e^{2x}$, multiplying the integrating factor by the original DE and simplifying gives:

$$e^{2x} y' + e^{2x} 2y = 0.$$
$$\underbrace{e^{2x} y' + e^{2x} 2y}_{\frac{\partial}{\partial x} e^{2x} y} = 0.$$

$$\text{Integrating gives, } e^{2x} y = \int 0.$$
$$e^{2x} y = c.$$
$$\text{Or, } y(x) = ce^{-2x}.$$

This equation is also a separable equation and can be solved as:

$$y' + 2y = 0.$$
$$\frac{dy}{dx} = -2y.$$
$$\frac{dy}{y} = -2dx.$$
$$\ln y = -2x + c.$$
$$Or: y(x) = ke^{-2x}, \text{ where } k = e^c$$

This solution agrees with the above solution.

17. $\quad y' + y = xy^4.$

Multiply the DE by y^{-4} gives,

$$y^{-4}y' + y^{-3} = x.$$

Choosing $u = y^{-3}$, then $du/dx = -3y^{-2}dy/dx$, then substitute back and simplify:

$$-\frac{1}{3}\frac{du}{dx} + u = x.$$
$$\frac{du}{dx} - 3u = -3x.$$

This is a linear DE with $p(x) = -3$, and integrating factor $I(x) = e^{\int p(x)dx} = \int e^{-3dx} = e^{-3x}$.
Multiplying this integrating factor by the linear equation above, and rearranging to get the solution:

$$e^{-3x}\frac{du}{dx} - e^{-3x}3u = -3xe^{-3x}.$$

2.10. REVIEW EXERCISE SOLUTIONS

$$\underbrace{e^{-3x}\frac{du}{dx} - e^{-3x}3u}_{\frac{d}{dx}(e^{-3x}u)} = -3xe^{-3x}.$$

$$\frac{d}{dx}(e^{-3x}u) = -3xe^{-3x}.$$

$$e^{-3x}u = -3\int xe^{-3x}dx.$$

$$e^{-3x}u = -3\,x(-\frac{1}{3})e^{-3x}$$
$$+ 3\int -1/3 e^{-3x}dx.$$

$$e^{-3x}u = xe^{-3x} - \int e^{-3x}dx.$$

$$e^{-3x}u = -3\,x(-\frac{1}{3})e^{-3x}$$
$$+ 3\int -1/3 e^{-3x}dx.$$

$$e^{-3x}u = xe^{-3x} + \frac{1}{3}e^{-3x} + c.$$

$$u = x + \frac{1}{3} + ce^{3x}.$$

$$\text{Or } y^{-3} = x + \frac{1}{3} + ce^{3x}.$$

18. $y' - y^2/x^2 = -y/x.$

Rearrange the equation as Bernoulli type:

$$y' + y/x = y^2/x^2.$$
$$y' + \frac{1}{x}y = \frac{1}{x^2}y^2.$$

Multiply both sides by y^{-2} to get rid of y^2 on the R.S of the equation:

$$y^{-2}y' + \frac{1}{x}y^{-1} = \frac{1}{x^2}.$$

Let $u = y^{-1}$, then $\frac{du}{dx} = -y^{-2}\frac{dy}{dx}$, substitute this back into the equation,

$$-\frac{du}{dx} + \frac{1}{x}u = \frac{1}{x^2}.$$

Or, $\quad \dfrac{du}{dx} - \dfrac{1}{x}u = -\dfrac{1}{x^2}.$

This is a Linear equation with $p(x) = -\frac{1}{x}$, and Integrating factor $I(x) = e^{\int p(x)dx} = e^{-\int \frac{dx}{x}} = \frac{1}{x}$. Multiplying by the integrating factor and simplifying to get the answer,

$$\frac{1}{x}\frac{du}{dx} - \frac{1}{x}\frac{1}{x}u = -\frac{1}{x}\frac{1}{x^2}.$$

$$\frac{1}{x}\frac{du}{dx} - \frac{1}{x^2}u = -\frac{1}{x^3}.$$

$$\underbrace{\frac{1}{x}\frac{du}{dx} - \frac{1}{x^2}u}_{\frac{d}{dx}(\frac{1}{x}u)} = -\frac{1}{x^3}.$$

$$\frac{d}{dx}\left(\frac{u}{x}\right) = -\frac{1}{x^3}.$$

$$\frac{u}{x} = -\int x^{-3}dx.$$

$$\frac{u}{x} = \frac{1}{2}x^{-2} + c.$$

$$u = \frac{1}{2}x^{-1} + cx^{-1}.$$

Or, $\quad y^{-1} = \dfrac{1}{2}x^{-1} + cx^{-1}.$

19. $\quad xy^2 y' + 2y^3 = 1.$

$$y' + \frac{2}{x}y = \frac{1}{x}y^{-2}.$$

2.10. REVIEW EXERCISE SOLUTIONS

$$y^{-2}y' + \frac{2}{x}y^{-1} = \frac{1}{x}.$$

Let $u = y^{-1}$, then $\frac{du}{dx} = -y^{-2}\frac{dy}{dx}$, substitute this back into the equation,

$$-\frac{du}{dx} + \frac{2}{x}u = \frac{1}{x}.$$

Or, $\quad \dfrac{du}{dx} - \dfrac{2}{x}u = -\dfrac{1}{x}.$

This is a Linear equation with $p(x) = -\frac{2}{x}$, and Integrating factor $I(x) = e^{\int p(x)dx} = e^{-2\int \frac{dx}{x}} = \frac{2}{x}$. Multiplying by the integrating factor and simplifying to get the answer,

$$\frac{2}{x}\frac{du}{dx} - \frac{2}{x^2}u = -\frac{2}{x^2}.$$

$$u = 1 + cx.$$

Or, $\quad y^{-1} = cx + 1.$

20. Solve the initial value problem:

$$\frac{dy}{dx} + \frac{x}{2y^2} = \frac{y}{x}; \quad y(1) = 2.$$

Rearranging the differential equation as,

$$\frac{dy}{dx} + \frac{x}{2}y^{-2} = \frac{1}{x}y.$$

The DE looks like Bernoulli type if rearranged in order,

$$\frac{dy}{dx} - \frac{1}{x}y = -\frac{x}{2}y^{-2}.$$

Multiplying both sides of the DE by y^2,
$$y^2\frac{dy}{dx} - \frac{1}{x}y^3 = -\frac{x}{2}.$$

As done previously, let $u = y^3$, then $\frac{du}{dx} = 3y^2\frac{dy}{dx}$, substitute this back into the equation,
$$\frac{1}{3}\frac{du}{dx} - \frac{1}{x}u = -\frac{1}{2}x.$$
$$\frac{du}{dx} - \frac{3}{x}u = -\frac{3}{2}x.$$

Which is a Linear equation with $p(x) = -\frac{3}{x}$, and integrating factor $I(x) = e^{-3\int \frac{dx}{x}} = \frac{1}{x^3}$. Multiplying this integrating by the equation and solve leads to the following solution:
$$y^3 = \frac{3}{2}x^2 + cx^{-3}.$$

To find the value of c we need to apply the initial values:
$$2^3 = \frac{3}{2}(1)^2 + c(1)^{-3}.$$

Then, $c = \frac{13}{2}$, and the general solution is:
$$y^3 = \frac{3}{2}x^2 + \frac{13}{2}x^{-3}.$$

21. The DE for the tank problem is, using the model formula for the system:
$$\frac{ds(t)}{dt} = input\ rate - output\ rate. \qquad (2.117)$$
$$\frac{ds(t)}{dt} = 2 - \frac{3s(t)}{(500+t)}. \qquad (2.118)$$

2.11 Chapter-2 Assignment

In the problems 1 - 3 solve the separable equations:

1. $y'' = t^3$.

2. $\dfrac{dy}{dx} = \dfrac{2x}{y-1}$.

3. $\dfrac{dy}{dx} = \dfrac{2}{x^2 y^2}$.

Solve the following exact equations:

4. $(y^2 + 2x)dx + (x^2 - 2y)dy = 0$.

5. $(2ty + 3)dt + t^2 dy = 0$.

6. $linxdt + \dfrac{t}{x}dx = 0$.

Solve the following equations:

7. $(x^5 - y)dx + xdy = 0$.

8. $dy - xdx = 0$.

9. $\dfrac{dy}{dx} + 3x^3 y = 0$.

10. $- e^{2x} y = 0$.

11. Solve the non-homogeneous equations:

$\dfrac{dy}{dx} + y = x$.

12. Find the general solution to the given DE:

$$\frac{dy}{dx} - 3y = 2e^{3x}.$$

In problems 13 - 14 Solve the IVP:

13. $x^4 \dfrac{dy}{dx} + 4x^3 y = x^2; \quad y(0) = 1.$

14. $\cos x \dfrac{dy}{dx} + y \sin x = x \cos x.$

Solve the following Tank-Problems:
15. Determine the amount of the salt in the tank of problem 22, if the outflow rate of the solution is increased to the rate of $6 gal/min$.

16. Do problem (22) with inflow rate=6 gal/min, and outflow rate=4 gal/min.

Chapter 3

Linear Second Order Equations

3.1 Introduction

First we shall consider a special physical situation "the Simple Pendulum" as an example from real life that leads to a second order linear differential equation, and shall present the theoretical means for obtaining the solution by applying Newton's second law, as was done in Chapter(2).

3.2 The Simple Pendulum

The apparatus of the Simple Pendulum consists of a small object (ball) of mass (m) attached to a cable of length (l) with negligible weight. Letting the ball swing in a vertical plane. Find the linear equation of the motion at any time. Applying Newton's Second law $F = ma$, where F is the total vector forces acting on the object of mass m and acceleration a, first, we need to analyze the vectors that are acting on the mass m shown in $fig(4)$.

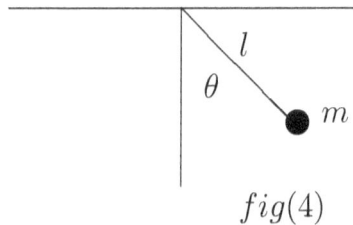

$fig(4)$

Now we shall study the forces(vectors) that are acting on the mass m as shown in fig(5) below:

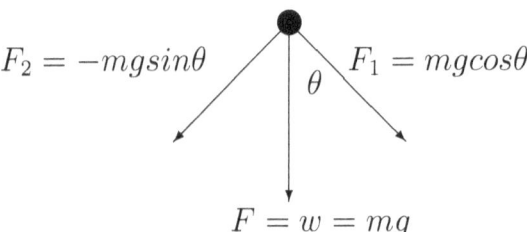

$fig(5)$ $vectors$ $acting$ on $mass$ m

The force F represents the weight of the object $w = mg$ pointing downward, this force is resolved into two (forces) components:

$$F_1 = mgcos\theta. \qquad (3.1)$$

Where θ is the angular displacement, and ,

$$F_2 = -mgsin\theta, \qquad (3.2)$$

3.2. THE SIMPLE PENDULUM

which is tangent to the circular path that the pendulum is swinging in. Letting the circular displacement be $s = l\theta$, then applying Newton's second law gives,

$$m\frac{d^2s}{dt^2} = F_2 = -mg\sin\theta. \tag{3.3}$$

$$\frac{d^2s}{dt^2} = -g\sin\theta. \tag{3.4}$$

$$l\frac{d^2\theta}{dt^2} + g\sin\theta = 0. \tag{3.5}$$

$$\frac{d^2\theta}{dt^2} + \frac{g}{l}\sin\theta = 0. \tag{3.6}$$

Equation (3.6) is a non-linear equation that describes the motion of a Simple Pendulum. If the initial angular displacement and acceleration is given, the question then can be stated as: Solve the non-linear second order homogeneous differential equation IVP:

$$\frac{d^2\theta}{dt^2} + \frac{g}{l}\sin\theta = 0. \tag{3.7}$$

With initial conditions:

$$\theta(0) = a. \tag{3.8}$$
$$\theta'(0) = b. \tag{3.9}$$

Where, a, and b are constants.

It is not easy to find the solution to a non-linear equation. Sometimes a non-linear equation is approximated by a linear equation, and this is called "Linearization" of the equation.

To solve the non-linear equation (3.7) means to find the value of $\theta(t)$.

If we try to observe the motion of the pendulum at a very small displacement (θ), then from Calculus we can assume that $(\sin\theta \cong \theta)$, and this will reduce equation (3.7) into:

$$\frac{d^2\theta}{dt^2} + \frac{g}{l}\theta = 0. \qquad (3.10)$$

Equation (3.10) is a second order linear-equation in θ. To solve this equation the best solution will be a solution of the form:
$\theta(t) = \cos(wt)$, or $\theta(t) = \sin(wt)$. Suppose we choose the first solution: $\theta(t) = \cos(wt)$ differentiating twice and substituting in equation (3.7) gives,

$$\theta(t) = \cos(wt). \qquad (3.11)$$
$$\theta'(t) = -wsis(wt). \qquad (3.12)$$
$$\theta''(t) = -w^2\cos(wt). \qquad (3.13)$$
$$-w^2\cos(wt) + \frac{g}{l}\cos(wt) = 0. \qquad (3.14)$$
$$w = \pm\sqrt{\frac{g}{l}}. \qquad (3.15)$$

In a similar way choosing a solution as $\theta(t) = \sin(wt)$ gives the same result. Since both $\cos(wt)$ and $\sin(wt)$ proves to be solutions to the linear differential second order equation, Then the general solution to the linear differential equation in (3.7) can be written as:

$$\theta(t) = C_1\cos(wt) + C_2\sin(wt). \qquad (3.16)$$

Where, C_1, and C_2 are constants and can be found by applying the initial conditions.

3.3 The Wronskian for 2nd ODE

As stated previously Linear differential equations are subdivided into two groups:
- Homogeneous Linear equations.
- Non-Homogeneous Linear equations.

The standard form for the linear non-homogeneous differential equations can be written as:

$$\frac{d^2y}{dx^2} + P(x)\frac{dy}{dx} + Q(x)y = G(x). \qquad (3.17)$$

Letting $G(x) = 0$ will reduce equation (3.17) into a Homogeneous equation,

$$\frac{d^2y}{dx^2} + P(x)\frac{dy}{dx} + Q(x)y = 0. \qquad (3.18)$$

To solve (3.18) we need to apply the Wronskian Method.

<u>The Wronskian:</u>
The Wronskian was named after the polish Mathematician (1778 − 1853). The Wronskian can be written in determinant form.

Definition: For any two functions $y_1(x)$, and $y_2(x)$, the Wronskian function is:

$$\begin{aligned}\mathbf{W(y_1, y_2)} &= \begin{bmatrix} y_1 & y_2 \\ y_1' & y_2' \end{bmatrix} \\ &= y_1 y_2' - y_2 y_1'.\end{aligned}$$

Applying the product rule of differentiation, we can get the first derivative of the Wronskian as follows:

$$\begin{aligned}\mathbf{W'(y_1, y_2)} &= y_1 y_2'' + y_1' y_2' - y_2' y_1' - y_2 y_1'' \\ &= y_1 y_2'' - y_2 y_1''.\end{aligned}$$

Wronskian = (1st function x derivative of the second function) - (second function x derivative of the first function). and,
d(Wronskian) = (first function x second derivative of the second function) - (second function) x (second derivative of the first function).

The Wronskian for n functions:
If $y_1, y_2, \ldots, y_n$ are solutions in some interval (a, b) for the differential equation:

$$y^n(x) + p_1(x)y^{n-1}(x) + \ldots + p_n(x)y(x) = 0 \quad (3.19)$$

Where, $p_1, p_2, \ldots, p_n$ are constants on (a, b) for x_0 in (a, b). then the Wronskian is:

$$W[\mathbf{y_1}, \ldots, \mathbf{y_n}](\mathbf{x_0}) = \begin{bmatrix} y_1 & y_2 & ,\ldots, y_n \\ y_1' & y_2' & ,\ldots, y_n' \\ \vdots & & \\ y_1^{n-1} & y_2^{n-1} & ,\ldots, y_n^{n-1} \end{bmatrix}$$

Then the solution on (a, b) can be expressed as:

$$y(x) = c_1 y_1 + c_2 y_2 + \ldots + c_n y_n.$$

Where, $c_1, c_2, \ldots c_n$ are arbitrary constants.

Properties of the Wronskian

For the homogeneous second order differential equations:

$$y'' + P(x)y' + Q(x)y = 0. \quad (3.20)$$

With two solutions: $y_1(x)$, and $y_2(x)$ in the interval (a, b)

3.3. THE WRONSKIAN FOR 2ND ODE

1. The Wronskian $W[y_1, y_2] = 0$ if and only if the solutions y_1, and y_2 are **linearly dependent**, then the solution of the differential equation can be expressed as:

$$c_1 y_1(x) + c_2 y_2(x) = 0.$$

Where, c_1, and c_2 are constants not both equal zero.

2. The Wronskian $W[y_1, y_2] \neq 0$ if and only if the solutions y_1, and y_2 are **linearly independent**, then the solution of the differential equation is called the particular solution $y_p(x)$ or the fundamental solution, then the general solution is given as:

$$y(x) = c_1 y_1(x) + c_2 y_2(x).$$

Where c_1, and c_2 are arbitrary constants.

Remark: In some cases it is possible that two differential functions are linearly independent, but their Wronskian equal 0. For example:

$$y_1 = x^2.$$
$$y_2 = |x^2|.$$
$$W(y_1, y_2) = \begin{bmatrix} x^2 & |x^2| \\ 2x & |2x| \end{bmatrix} = 0.$$

Even though y_1 is not a constant multiple of the other function for all x in the interval. Also, the reason is because y_1, and y_2 are not the solutions for the same homogeneous equation in the interval.

Remark: Another form of Wronskian is called "**Abel's Identity**", and was derived in 1827. This identity states that: The Wronskian for two solutions y_1, and y_2 to the equation:

$$y'' + py' + qy = 0.$$

Is given as:

$$W[y_1, y_2](x) = ce^{-\int_{x_0}^{x} p(t)dt} = \frac{c}{I(x)]_{x_0}^{x}}.$$

Where, c is the constant that depends on y_1, and y_2, and $I(x)$ is the integrating factor. This identity confirms that $W = 0$ for $c = 0$, and $W \neq 0$ for $c \neq 0$.

Example - 1: Use Abel's Identity to determine the Wronskian of two solutions on $(0, \infty)$ to the differential equations:

$$xy'' + (x-1)y' + 3y = 0.$$
$$y'' + \frac{x-1}{x}y' + \frac{3}{x}y = 0.$$

Solution:
Their Wronskian is:

$$\begin{aligned} W &= ce^{-\int p(x)dx} \\ &= ce^{-\int \frac{x-1}{x} dx} \\ &= ce^{-x+\ln x}. \\ Then, \ W &= cxe^{-x}. \end{aligned}$$

3.3. THE WRONSKIAN FOR 2ND ODE

Example - 2: Linearly Dependent Solution
Determine whether the functions :
$y_1(t) = 2cost$, and $y_2(t) = 0$, are linearly dependent.

Solution:
By computing their Wronskian:

$$W(y_1, y_2) = W(2cost, 0).$$
$$= \begin{bmatrix} 2cost & 0 \\ -2sint & 0 \end{bmatrix}$$
$$= 0.$$

Then the two solutions are linearly dependent and the general solution is : $y(t) = c_1 y_1 + c_2 y_2 = 2c_1 cost$.

Example - 3: Linearly Independent Solution
Determine whether the functions: $y_1(t) = 2sint$, and $y_2(t) = 3cost$, are linearly independent.

Solution:
By computing their Wronskian:

$$W(y_1, y_2) = W(2sint, 3cost).$$
$$= \begin{bmatrix} 2sint & 3cost \\ 2costt & -3sint \end{bmatrix}$$
$$= -6 \neq 0.$$

Then the two solutions are linearly independent and the general solution is called the particular solution $y_p(t)$.

Existence and Uniqueness of the solution for second order IVP

For the second order differential equation,

$$\frac{d^2y}{dx^2} + P(x)\frac{dy}{dx} + Q(x)y = G(x). \quad (3.21)$$

With the initial conditions,

$$y(x_0) = y_0, \quad (3.22)$$
$$y(x_0) = y_1. \quad (3.23)$$

Where, $P(x)$, $Q(x)$, and $G(x)$ are continuous functions on an open interval $a < x < b$. Then for any choice of initial values, y_0, y_1, there exist a unique solution $y(x)$ for the initial value problem.

Remark: The uniqueness for two solutions means the intersection at one point in the interval.

Example - 4:
Determine the existence and uniqueness of the solution to the initial value problem,

$$(t-4)y'' + y' + (t-4)\sqrt{t}y = (t-4)\ln(t). \quad (3.24)$$
$$y(1) = 4, \quad (3.25)$$
$$y'(1) = -6. \quad (3.26)$$

Solution:
To rewrite the differential equation in the standard form as in (3.20), we shall divide each term by the coefficient of y'' i.e $(t-4)$,

$$y'' + \frac{1}{t-4}y' + \sqrt{t}y = \ln(t). \quad (3.27)$$

From equation (3.27) we get:
$$P(t) = \frac{1}{t-4}.$$
$$Q(t) = \sqrt{t}.$$
$$G(t) = \ln(t).$$

Where, $P(t)$ is continuous for $t \neq 4$, $Q(t)$ is continuous for $t > 0$, and $G(t)$ is continuous for $t > 0$.
Then we conclude that the IVP has a unique solution in the open interval $(0, 4)$.

3.4 Finding a Second Solution from a Given One

Method of Reduction of Order (1) D'Alemberts Method

If a homogeneous differential equation of second order is given,
$$a_2(x)y'' + a_1(x)y' + a_0(x)y = 0. \qquad (3.28)$$
Where, $a_2(x)$, $a_1(x)$, and $a_0(x)$ are variable coefficients. If one solution $y_1(x)$ is given, then a second linearly independent solution $y_2(x)$ can be found in the following steps:

Step-1:
Rewrite equation (3.28) in the standard form, which requires the division by the leading coefficient $a_2(x)$, then letting, $\frac{a_1(x)}{a_2(x)} = P(x)$, and $\frac{a_0(x)}{a_2(x)} = Q(x)$, where $P(x)$, and $Q(x)$ are variable functions.
$$y'' + P(x)y' + Q(x)y = 0. \qquad (3.29)$$

Step-2:
For simplicity let the given solution $y_1(x) = q$, and Choose a second solution to be :

$$y_2(x) = uq, \qquad (3.30)$$

where $y_2(x)$ is linearly independent of $y_1(x)$, and $u = u(x)$.

Step-3:
Differentiate equation (3.30) twice, with respect to x:

$$\begin{aligned}
y_2'(x) &= u'q + uq'. & (3.31) \\
y_2''(x) &= u''q + u'q' + u'q' + uq''. & (3.32) \\
&= u''q + 2u'q' + uq''. & (3.33)
\end{aligned}$$

Substitute back into equation (3.29) and rearrange:

$$u''q + 2u'q' + uq'' + Pu'q + Puq' + Quq = 0.$$
$$u(q'' + Pq' + Qq) + u''q + u'(2q' + Pq) = 0.$$

The first term in between parenthesis is just equation (3.29) and is equal zero. Then the equation reduces to :

$$qu'' + (2q' + Pq)u' = 0. \qquad (3.34)$$

Step-4:
Now, we need to solve equation (3.34). Let $u' = v$, then $u'' = v'$, and substitute back into equation (3.34) gives,

$$qv' + (2q' + Pq)v = 0. \qquad (3.35)$$

This equation is a linear equation and separable, it can be separated as follows,

$$\frac{v'}{v} = -\frac{2q' + q}{q} = -\frac{2q'}{q} - P.$$

$$\frac{dv}{v} = -2\int \frac{dq}{q} - \int Pdx.$$

$$lnv = -2lnq - \int Pdx.$$

$$Or, v = e^{q^{-2} - \int Pdx}.$$

$$= \frac{e^{-\int Pdx}}{q^2}.$$

But $v = u'(x)$, then,

$$u(x) = \int \frac{e^{-\int Pdx}}{q^2} dx. \qquad (3.36)$$

Substituting back into equation (3.30) gives the second linearly independent solution to the differential equation in (3.29).

$$y_2(x) = u(x)y_1(x) = y_1(x) \int \frac{e^{-\int Pdx}}{y_1^2} dx \qquad (3.37)$$

Thus in solving a differential equation when one solution is provided and to avoid going through all these steps to find a second solution, we can simply apply the formula in (3.37) directly. as its illustrated in the following example.

Remark - 1: The method of deriving equation (3.37) is called: **The Method of Reduction of Order**, also known as **D'Alembert's Method**. It was discovered by Jean D'Alemberts (1717-1783).

Remark - 2: Notice in equation (3.37) the numerator $e^{-\int Pdx}$ is the integrating factor $I(x)$, then equation (3.37) can also be expressed as ,

$$second\ solution = first\ solution \int \frac{Integrating\ factor}{(first\ solution)^2}. \qquad (3.38)$$

Example - 5:
Find a second linearly independent solution for the differential equation:

$$x^3 y'' - 3x^2 y' + 4xy = 0. \qquad (3.39)$$

If one solution is given as $y_1(x) = x^3$.

Solution:
Rewriting the equation in the standard form,

$$y'' - \frac{3}{x} y' + \frac{4}{x^2} y = 0 \qquad (3.40)$$

Where, $P(x) = -3/x$. Then the integrating factor for the differential equation is:

$$\begin{aligned} I(x) &= e^{-\int P(x)dx} \\ &= e^{-\int \frac{-3}{x} dx} \\ &= e^{3\ln x} \\ &= x^3. \end{aligned}$$

Applying formula (3.36) or (3.37) directly, we get the second linearly independent solution $y_2(x)$,

$$y_2(x) = x^3 \int \frac{x^3}{x^6} dx. \qquad (3.41)$$

$$y_2(x) = x^3 \int x^{-3} dx. \qquad (3.42)$$

$$Then, \ y_2(x) = -\frac{1}{2} x + c. \qquad (3.43)$$

3.5 Method of Reduction of Order(2)

When dependent variable (y) is missing

Example - 6 : Homogeneous DE Missing y
Solve the homogeneous differential equation:
$$y'' + (y')^2 = 0. \tag{3.44}$$

Solution:
Since the dependent variable y is missing, then we can reduce the order of the equation with the following substitution: Let $y' = u$, then $y'' = u'$, then equation (3.44) will be,
$$u' + u^2 = 0. \tag{3.45}$$
This separable equation can be written and simplify as,

$$\frac{du}{dx} + u^2 = 0.$$
$$\frac{du}{dx} = -u^2.$$
$$\frac{du}{u^2} = dx.$$
$$u^{-2} du = dx.$$
$$\int u^{-2} du = \int dx.$$
$$-\frac{1}{u} = -x + c.$$
$$\frac{1}{u} = -c = x.$$
$$\frac{1}{u} = k + x. \quad where, \ k = -c.$$
$$Then, \ u = \frac{1}{k+x}.$$

But $u = y'$, then replacing u with y' and solve to get the final solution.

$$u = y' = \frac{1}{k+x}.$$
$$\frac{dy}{dx} = \frac{1}{k+x}.$$
$$\int dy = \int \frac{dx}{k+x}.$$
$$Then, y(x) = ln(k+x) + c.$$

Example - 7 : Non-Homogeneous DE Missing y
Solve the non-homogeneous differential equation:

$$xy'' + y' = x^2. \qquad (3.46)$$

In the same manner as in example-1: Since the dependent variable y is missing, then we can reduce the order of the equation with the following substitution: Let

$$y' = u.$$
$$Then \ y'' = u'.$$

And, equation (3.46) becomes,

$$xu' + u = x^2.$$
$$Or, u' + \frac{1}{x}u = x$$

This equation is a standard form linear equation with $P(x) = \frac{1}{x}$, and integrating factor $I(x) = e^{\int p(x)} = e^{\int \frac{dx}{x}} = e^{lnx} = x$. Then multiply this standard equation by the integrating factor x,

$$xu' + u = x^2.$$
$$\underbrace{xu' + u}_{\frac{d}{dx}(xu)} = x^2.$$

Then, integrating both sides gives,

$$xu = \int x^2 dx.$$

$$xu = \frac{x^3}{3} + c.$$

$$\text{Then, } u = \frac{x^3}{3} + \frac{c}{x}.$$

$$\text{Or, } u = y' = \frac{x^2}{3} + \frac{c}{x}.$$

$$y(x) = \int \frac{x^2}{3} dx + \int \frac{c}{x} dx.$$

$$y(x) = \frac{x^3}{9} + c \ln x + k.$$

3.6 Homogeneous Equations with Constant Coefficients

We have studied the homogeneous differential equation of second order with coefficients as functions of the dependent variable as given in the following equation,

$$a_2(x)y'' + a_1(x)y' + a_0(x)y = 0. \qquad (3.47)$$

Where, $a_2(x)$, $a_1(x)$, and $a_0(x)$ are coefficients and functions of the independent variable x. Letting $a_2(x) = a_2$, $a_1(x) = a_1$, and $a_0(x) = a_0$, and writing equation (3.47) as:

$$a_2 y'' + a_1 y' + a_0 y = 0. \qquad (3.48)$$

Where, a_2, a_1, and a_0 are all constants, then equation (3.48) is a second order homogeneous equation with constant coefficients. Since the equation is of degree -2 then two solutions are expected say: y_1, and y_2. If these two solutions are linearly independent which means their Wronskian :

$$W(y_1, y_2) = \begin{bmatrix} y_1 & y_2 \\ y_1' & y_2' \end{bmatrix} \neq 0.$$

Then the general solution for the differential equation can be written as:

$$y_G = c_1 y_1 + c_2 y_2. \qquad (3.49)$$

A solution of type $y = e^{rt}$ is used to solve these equations.

$$\begin{aligned} y &= e^{rt}. \\ y' &= re^{rt}. \\ y'' &= r^2 e^{rt}. \end{aligned}$$

Substitute back into equation (3.48) gives.

$$a_2 r^2 + a_1 r + a_0 = 0. \qquad (3.50)$$

This equation is called the Auxiliary equation, which resembles the quadratic equation, with $a_2 = a$, $a_1 = b$, and $a_0 = c$. To find the general solution , we find the two Linearly-Independent solutions y_1, and y_2. and this is done by treating equation (3.50) as a quadratic equation, and using the quadratic formula to solve it. The discriminant can be tested for the expected type of solutions :

1. if the discriminant $(b^2 - 4ac) > 0$, then the two solutions y_1, and y_2 are real and distinct solutions, and their general solution will be:

$$\begin{aligned} y &= c_1 y_1 + c_2 y_2. \\ y &= c_1 e^{r_1 x} + c_2 e^{r_2 x}. \text{ where } r_1 \neq r_2. \end{aligned}$$

3.6. HOMOGENEOUS EQUATIONS WITH

2. if the discriminant $(b^2 - 4ac) = 0$, then the two solutions y_1, and y_2 are equal to each other (repeated), and their general solution will be:

$$y = c_1 y_1 + c_2 y_2.$$
$$y = c_1 e^{rx} + c_2 x e^{rx}. \text{ where } r_1 = r_2 = r.$$

3. if the discriminant $(b^2 - 4ac) < 0$, then the two solutions y_1, and y_2 are non-real and distinct solutions, and their general solution will be:

$$y = c_1 y_1 + c_2 y_2.$$
$$y = c_1 e^{(\alpha+i\beta)x} + c_2 e^{(\alpha-i\beta)x}.$$
$$y = e^{\alpha x} \{c_1 e^{i\beta x} + c_2 e^{-i\beta x}\}.$$
$$y = e^{\alpha x} \{c_1 cos\beta x + c_2 sin\beta x\}.$$

Example - 8 : Real and distinct roots
Find the general solution for the differential equation :

$$3y'' + 7y' + 2y = 0 \qquad (3.51)$$

Using the solution of the type $y = e^{rt}$, and differentiate twice:

$$y = e^{rt}.$$
$$y' = re^{rt}.$$
$$y'' = r^2 e^{rt}.$$

Then substituting into equation (3.51) gives the following auxiliary equation:

$$3r^2 + 7r + 2 = 0 \qquad (3.52)$$

Which has the following roots: $r_1 = -2$, and $r_2 = -\frac{1}{3}$.
Thus the general solution for the differential equation is:

$$y_G = c_1 y_1 + c_2 y_2.$$
$$y_G = c_1 e^{r_1 t} + c_2 e^{r_2 t}.$$
$$y_G = c_1 e^{-2t} + c_2 e^{-\frac{t}{3}}.$$

Example - 9 : Real and repeated roots
Find the general solution for the differential equation :

$$2y'' + 4y' + 2y = 0 \qquad (3.53)$$

Using the solution of the type $y = e^{rt}$, and differentiate twice:

$$y = e^{rt}.$$
$$y' = re^{rt}.$$
$$y'' = r^2 e^{rt}.$$

Then substituting into equation (3.53) gives the following auxiliary equation:

$$2r^2 + 4r + 2 = 0 \qquad (3.54)$$

Which has the following roots: $r_1 = r_2 = -1$. Thus the general solution for the differential equation is:

$$y_G = c_1 y_1 + c_2 y_2.$$
$$y_G = c_1 e^{r_1 t} + c_2 e^{r_2 t}.$$
$$y_G = c_1 e^{-t} + c_2 t e^{-t}.$$

3.6. HOMOGENEOUS EQUATIONS WITH   103

Example - 10 : Non-Real and distinct roots

Find the general solution for the differential equation :

$$y'' + y' + 3y = 0 \tag{3.55}$$

Using the solution of the type $y = e^{rt}$, and differentiate twice:

$$\begin{aligned} y &= e^{rt}. \\ y' &= re^{rt}. \\ y'' &= r^2 e^{rt}. \end{aligned}$$

Then substituting into equation (3.55) gives the following auxiliary equation:

$$r^2 + r + 3 = 0 \tag{3.56}$$

Which has the following roots:

$$\begin{aligned} r_{1,2} &= \frac{-1 \pm \sqrt{1 - 4(1)(3)}}{2(1)} \\ &= \frac{-1 \pm \sqrt{11}i}{2}. \end{aligned}$$

Thus the general solution for the differential equation is:

$$\begin{aligned} y_G &= c_1 y_1 + c_2 y_2. \\ y_G &= c_1 e^{r_1 t} + c_2 e^{r_2 t}. \\ y_G &= e^{-t/2} \{c_1 \cos \frac{\sqrt{11}}{2} t + c_2 \sin \frac{\sqrt{11}}{2} t\}. \end{aligned}$$

Remark: As mentioned before when the discriminant is $b^2 - 4ac < 0$, then the roots of the auxiliary equation:

$$ar^2 + br + c = 0,$$

are the complex conjugates:

$$r_1 = \alpha + i\beta.$$
$$r_2 = \alpha - i\beta.$$

And the general solution is:

$$y_G = c_1 e^{(\alpha+i\beta)x} + c_2 e^{(\alpha-i\beta)x}.$$
$$= e^{\alpha x}\{c_1 e^{i\beta x} + c_2 e^{-i\beta x}\}.$$
$$\text{Or, } y_G = e^{\alpha x}\{c_1 cos\beta x + c_2 sin\beta x\}.$$

Where, $e^{i\theta} = cos\theta + isin\theta$ which is known as Euler's Formula.

3.7 Non-Homogeneous Eq's with Constant Coefficients

The non-homogeneous second order linear differential equation has the form,

$$a_2 y''(x) + a_1 y'(x) + a_0 y(x) = g(x). \tag{3.57}$$

Where, a_2, a_1, and a_0 are constants. In the previous section equation (3.57) was solved with $g(x) = 0$, but now we have to solve it with $g(x) \neq 0$ where the general solution $y_G(x)$ will have the following form:

$$y_G(x) = y_H(x) + y_P(x). \tag{3.58}$$

Where $y_H(x)$ is the homogeneous solution for the differential equation (3.57) with $g(x) = 0$ which we have solved in the previous section as :

$$y_H(x) = c_1 y_1 + c_2 y_2. \qquad (3.59)$$

And the particular solution is the solution of the differential equation (3.57) with $g(x) \neq 0$, and this can be accomplished by the following methods:
1. By inspection.
2. By the method of undetermined coefficients.
3. By the method of variation of parameters.
These 3-methods will be illustrated by the following examples:

Example - 11 : By Inspection
Find the particular solution for the differential equation:

$$y'' + y - 6 = 3e^{2x}$$

Solution:
Since $g(x)$ is of exponential form, then by inspection we assume that the particular solution is : $y_p(x) = Ae^{2x}$, then differentiate twice:

$$y_p = Ae^{2x}.$$
$$y_p' = 2Ae^{2x}.$$
$$y_p'' = 4Ae^{2x}.$$

And substitute back into the differential equation gives:

$$4Ae^{2x} + Ae^{2x} - 6 = 3e^{2x}.$$

Solving for A gives, $A = \frac{3}{5}$, then the particular solution is: $y_P(x) = 3/5 e^{2x}$.

Method of Undetermined Coefficients

The method of undetermined coefficient for finding the particular solution $y_P(x)$ for the non-homogeneous linear equation with constant coefficients works when the non-homogeneous term $g(x)$ is one of the following:
1. Exponential functions.
2. Polynomial functions.
3. Trigonometric functions.

Or the product of the above three functions. In the following table we list examples that guide us in our strategy:

$g(x)$	y_P
3	A
2X	Ax+B
5 x^2	Ax^2+Bx+C
e^{3x}	Ae^{3x}
sinx	Acosx + B sinx
(3x+5) e^{2x}	(Ax+B)e^{2x}
x^2sinx	(Ax^2+Bx +C)sinx+(Dx^2+Ex+F)cosx

Table(1)

Example - 12 : by undetermined Coefficients

Find the particular solution for the differential equation:

$$y'' - y = 5x$$

The homogeneous solution is of the exponential type. And since $g(x) = 5x$ the particular solution will be as shown on the table: $y_p = Ax + B$. Differentiating y_P twice gives:

$$y_P = Ax + B.$$

$$y'_P = A.$$
$$y''_P = 0.$$

Substituting back into differential equation and solve for the constants A, and B, gives: $A = -5$, and $B = 0$, then the particular solution is $y_P = -5x$.

Example - 13: by Superposition Method

Solve the non-homogeneous equation using the superposition method:

$$u'' - u = x^2 - e^x + 3\sin x.$$

To solve this equation using the superposition method means to split the problem into 3-equations:

$$u'' - u = x^2.$$
$$u'' - u = -e^x.$$
$$u'' - u = 3\sin x.$$

For these three equation, we solve the homogeneous equation first.

$$u'' - u = 0.$$

Its auxiliary equation is,

$$r^2 - 1 = 0, then\ r = \pm 1.$$

And, the homogeneous solution is:

$$u_H = c_1 e^x + c_2 e^{-x}.$$

Solving the particular solution for : $u'' - u = x^2$, we choose :

$$u_{p1} = x^s(Ax^2 + Bx + C),\ where\ s = 0.$$

Then,
$$u_{p1} = Ax^2 + Bx + C.$$
$$u'_{p1} = 2Ax + B.$$
$$u''_{p1} = 2A.$$

Substituting back into the first equation,
$$u'' - u = x^2.$$
$$2A - Ax^2 - Bx - C = x^2.$$

Solving for the constants we get: $A = -1$, $B = 0$, and $C = 2$. Then:
$$u_{p1} = -x^2 + 2.$$

In a similar way, we let:
$$u_{p2} = x^s(ke^x). \ Where \ s = 1.$$
$$Then, \ u_{p2} = kxe^x.$$
$$u'_{p2} = ke^x + kxe^x.$$
$$u''_{p2} = 2ke^x + kxe^x.$$

Substituting back in to the second equation, and solving for the constant we get $A = -1/2$ and:
$$u_{p2} = -\frac{1}{2}xe^x.$$

In a similar way, we let:
$$u_{p3} = x^s(A sinx + B cosx), \ where \ s = 0.$$
$$Then \ u_{p3} = A sinx + B cosx.$$
$$u'_{p3} = A cosx - B sinx.$$
$$u''_{p3} = -A sinx - B cosx.$$

Substituting back into the third equation, and solving for the constant we get $A = -3/2$ and $B = 0$:

$$u_{p3} = -\frac{3}{2}sinx.$$

Then by the superposition method, the general solution for the problem is:

$$u_G = u_H + u_{p1} + u_{p2} + u_{p3}.$$
$$u_G = c_1 e^x + c_2 e^{-x} - x^2 + 2 - \frac{1}{2}xe^x - \frac{3}{2}sinx.$$

Note: s represents number of times the chosen particular solution appears in the homogeneous solution. Thus in choosing the particular solution it is better if we let $y_p = x^s g(x)$ this will give the correct solution to the problem.

Method of Variation of Parameters

The method of variation of parameters also known as **Lagrang's Method**, was derived by J. L. Lagrange (1736 − 1813) in 1774. The method is a procedure for solving linear equation which is particularly useful for higher order linear equations. The method is based on the idea that just knowing the form of the solution, we can substitute it into the given equation and solve for any unknown.

Variation of Parameters for Non-Homogeneous second ODE

Consider the second order differential equation:

$$y'' + p(x)y' + q(x)y = G(x). \tag{3.60}$$

The general solution y_G of this equation requires the solution of the homogeneous problem called the fundamental solution or y_H, and the non-homogeneous problem called the particular solution:

$$y_G = y_H + y_p. \tag{3.61}$$

The Homogeneous solution was found in the previous sections to be,

$$y_H = c_1 y_1 + c_2 y_2.$$

Here, our task is to find the non-homogeneous solution, known as the particular solution y_p, using the method of Variation of Parameters. Let the particular solution be,

$$y_p = u_1(x)y_1(x) + u_2(x)y_2(x). \tag{3.62}$$
$$Or,\ y_p = u_1 y_1 + u_2 y_2. \tag{3.63}$$

Differentiating (3.63) twice gives,

$$y'_p = u'_1 y_1 + u_1 y'_1 + u'_2 y_2 + u_2 y'_2,\ and$$
$$y''_p = u''_1 y_1 + u'_1 y'_1 + u'_1 y'_1 + u_1 y''_1 + u''_2 y_2$$
$$\quad + u'_2 y'_2 + u'_2 y'_2 + u_2 y''_2.$$

Substituting back into equation (3.60), and rearranging leads to a system of two equations:

$$G(x) = u'_1 y'_1 + u_1 y''_1 + u'_2 y'_2 + u_2 y''_2 + p(u_1 y'_1 + u_2 y'_2)$$
$$\quad + q(u_1 y_1 + u_2 y_2).$$
$$G(x) = u_1(y''_1 + py'_1 + qy_1) + u_2(y''_2 + py'_2 + qy_2)$$
$$\quad + (u'_1 y'_1 + u'_2 y'_2).$$

Since y_1, and y_2 are solutions to the homogeneous problem, then we set:
$$y_1'' + py_1' + qy_1 = 0,$$
$$\text{and,} \quad y_2'' + py_2' + qy_2 = 0.$$

Then, this leads to a system of two equations:
$$u_1'y_1 + u_2'y_2 = 0. \tag{3.64}$$
$$u_1'y_1' + u_2'y_2' = G(x). \tag{3.65}$$

To solve this system we use Cramer's Rule to get u_1, and u_2 as follows:

$$u_1' = \frac{\begin{bmatrix} 0 & y_2 \\ y & y_2' \end{bmatrix}}{\begin{bmatrix} y_1 & y_2 \\ y_1' & y_2' \end{bmatrix}}$$

$\Longrightarrow$

$$u_1 = \int \frac{-Gy_2}{W[y_1, y_2]} dx. \tag{3.66}$$

And,

$$u_2' = \frac{\begin{bmatrix} y_1 & 0 \\ y_2' & y \end{bmatrix}}{\begin{bmatrix} y_1 & y_2 \\ y_1' & y_2' \end{bmatrix}}$$

$\Longrightarrow$

$$u_2 = \int \frac{Gy_1}{W[y_1, y_2]} dx. \tag{3.67}$$

Example - 14: Variation of Parameters

This example will illustrate the steps of solving differential equation of second order using the method of variation of parameters. Find the general solution for the second order differential equation :

$$y'' + 9y = tan3x. \qquad (3.68)$$

Step-1: Solve the homogeneous problem y_H:

$$y'' + 9y = 0. \qquad (3.69)$$

This equation has a solution of the form $y = e^{rx}$.

$$y = e^{rx}.$$
$$y' = re^{rx}.$$
$$y'' = r^2 e^{rx}.$$

Substituting back into (3.73) gives,

$$r^2 + 9 = 0 \Longrightarrow r = \pm 3i.$$

Then writing the solution as:

$$\begin{aligned} y_H &= c_1 y_1 + c_2 y_2. \\ &= c_1 e^{r_1 x} + c_2 e^{r_2 x}. \\ &= c_1 e^{3i} + c_2 e^{-3i}. \\ \text{Then, } y_H &= c_1 cos3x + c_2 sin3x. \end{aligned}$$

Step-2: Solve the non-homogeneous problem y_p.

$$\text{Set } y_p = u_1 cos3x + u_2 sin3x.$$

Differentiating twice and substituting back into the original non-homogeneous equation, leads to the following system of two equations:

$$u_1' cos3x + u_2' sin3x = 0. \qquad (3.70)$$
$$-3u_1' sin3x + 3u_2' cos3x = tan3x. \qquad (3.71)$$

The Wronskian for the system is:

$$W(y_1, y_2) = W(cos3x, sin3x) = \begin{bmatrix} cos3x & sin3x \\ -3sin3x & 3cos3x \end{bmatrix}$$
$$= 3 \neq 0.$$

Then using the formulas derived by Cramer's rule we get:

$$u_1 = \int \frac{-tan3x\, sin3x}{3} dx.$$

$$\text{And, } u_2 = \int \frac{tan3x\, cos3x}{3} dx.$$

Then solving u_1, and u_2 gives,

$$\begin{aligned} u_1 &= -\frac{1}{3} \int sin3x\, tan3x dx. \\ &= -\frac{1}{3} \int sin3x\, \frac{sin3x}{cos3x} dx. \\ &= -\frac{1}{3} \int \frac{sin^2 3x}{cos3x} dx. \\ &= -\frac{1}{3} \int \frac{1 - cos^2 3x}{cos3x} dx. \\ &= \frac{1}{3} \int \frac{cos^2 3x}{cos3x} - \frac{1}{3} \int \frac{1}{cos3x} dx. \\ &= \frac{1}{3} \int cos3x dx - \frac{1}{3} \int sec3x dx. \end{aligned}$$

Then, $u_1 = \frac{1}{9}\{sin3x - ln(sec3x + tan3x)\} + c_1.$

And,

$$\begin{aligned} u_2 &= \frac{1}{3} \int cos3x.tan3x dx. \\ &= \frac{1}{3} \int cos3x.\frac{sin3x}{cos3x} dx. \end{aligned}$$

$$= \frac{1}{3}\int sin3x dx.$$

$$\text{Then, } u_2 = -\frac{1}{9}cos3x + c_2.$$

Letting c_1, and c_2 equal zero, then,

$$y_p = \frac{1}{9}\{sin3x - ln(sec3x + tan3x)\, cos3x + cos3x.sin3x\}.$$

$$\text{Or, } y_p = -\frac{1}{9}cos3x\, ln(sec3x + tan3x).$$

Step-3: Writing the general solution for the given differential equation (3.68):

$$y_G(x) = y_H(x) + y_p(x).$$

$$y_G = c_1 cos3x + c_2 sin3x - \frac{1}{9}cos3x\, ln(sec3x + tan3x).$$

Example-15 : Variation of Parameters

Find the general solution for the second order differential equation :

$$4y'' - 4y' + y = x^{-2}e^{x/2}. \tag{3.72}$$

Step-1: Solve the homogeneous problem y_H:

$$4y'' - 4y' + y = 0. \tag{3.73}$$

Since this is a second order equation with constant coefficients, the the solution has a form of $y = e^{rx}$, differentiating twice and substituting into y_H gives the following auxiliary equation:

$$4r^2 - 4r + 1 = 0$$

using quadratic formula, the auxiliary equation gives two solutions which form the homogeneous solution:

$$y_1 = y_2 = 1/2.$$
$$\text{And } y_H = c_1 e^{x/2} + c_2 x e^{x/2}.$$

Step-2: Solve the non-homogeneous problem y_p.

$$\text{Set } y_p = u_1 y_1 + u_2 y_2.$$

For simplicity let $u_1 = u$, and $u_2 = v$, and use the solutions from the homogeneous problem: $y_1 = e^{x/2}$, and $y_2 = x e^{x/2}$ to write y_p:

$$y_p = u y_1 + v y_2.$$

Differentiating twice and substituting back to the non-homogeneous equation, then rearranging to get the following system of two equations:

$$y_1 u' + y_2 v' = 0.$$
$$y_1' u' + y_2' v' = g(x).$$

Or,

$$e^{x/2} u' + x e^{x/2} v' = 0.$$
$$1/2 e^{x/2} u' + (x/2 e^{x/2} + e^{x/2}) v' = x^{-2} e^{x/2}$$

Then we find the Wronskian, and use Cramer's rules to solve for u, and v:

$$W(y_1, y_2) = W(e^{x/2}, x e^{x/2})$$
$$= \begin{bmatrix} e^{x/2} & x e^{x/2} \\ 1/2 e^{x/2} & x?2 e^{x/2} + e^{x/2} \end{bmatrix}$$
$$= e^x \neq 0.$$

Then u, and v are:

$$u = \int \frac{-x^{-2}e^{x/2} \; xe^{x/2}}{e^x} \, dx = -lnx.$$

$$v = \int \frac{x^{-2}e^{x/2} \; e^{x/2}}{e^x} \, dx$$

Taking the constant of integrations to be zero. Substituting these solutions into y_p gives:

$$y_p = -lnxe^{x/2} - e^{x/2} = -e^{x/2}(lnx + 1).$$

Then, the general solution for the problem is:

$$y_G = c_1 e^{x/2} + c_2 xe^{x/2} - e^{x/2}(lnx + 1).$$

3.8 A Special Case of Variable Coefficients

Cauchy - Euler Equations

There are special type of equations with variable coefficients, such as the following second order equation:

$$a_2 x^2 y'' + a_1 x y' + a_0 y = g(x). \; x > 0. \quad (3.74)$$

Where, a_2, a_1, and a_0 are constants, and the coefficients x^2, and x are not constant. Equation (3.74) can be transformed, by means of exponential substitution followed by a chain rule, into a second order equation with constant coefficients.

$$a_2 y'' + (a_1 - a_2) y' + a_0 y = g(x). \quad (3.75)$$

Such equation as (3.74) is called Cauchy - Euler equation.

Remark: This work was published by Leonard Euler in 1769, and later by Augustine Cauchy, that is why is called Cauchy - Euler.

The steps to transform equation (3.74) into equation (3.75) are shown as follows:

Step-1: Make the substitution $x = e^t$, and differentiate with respect to t.

$$x = e^t. \tag{3.76}$$

$$\frac{dx}{dt} = e^t \tag{3.77}$$

Then, differentiate y with respect to t using chain rule, to find xy'.

$$\frac{dy}{dt} = \frac{dy}{dx}\frac{dx}{dt}.$$

$$= \frac{dy}{dx}e^t.$$

$$\text{Or,} \quad \frac{dy}{dt} = x\frac{dy}{dx} = xy'.$$

Letting $\frac{dy}{dt} = y^\bullet$, then:

$$xy' = y^\bullet. \tag{3.78}$$

Differentiating again with respect to t using the chain rule to get $x^2 y''$.

$$\frac{d^2y}{dt^2} = \frac{d}{dt}(\frac{dy}{dt}).$$

$$= \frac{d}{dt}(x\frac{dy}{dx}).$$

$$\frac{d^2y}{dt^2} = \frac{dx}{dt}\frac{dy}{dx} + x\frac{d}{dt}(\frac{dy}{dx}).$$

$$= \frac{dy}{dt} + x\frac{d^2y}{dx^2}\frac{dx}{dt}.$$

$$= \frac{dy}{dt} + x\frac{d^2y}{dx^2}e^t.$$

$$= \frac{dy}{dt} + x\frac{d^2y}{dx^2}x.$$

$$= \frac{dy}{dt} + x^2\frac{d^2y}{dx^2}.$$

$$\text{Or, } x^2\frac{d^2y}{dx^2} = \frac{d^2y}{dt^2} - \frac{dy}{dt} = x^2y''.$$

Letting $\frac{d^2y}{dt^2} = y^{\bullet\bullet}$, then:

$$x^2y'' = y^{\bullet\bullet} - y^{\bullet}. \tag{3.79}$$

In the same manner using the chain rule we can show that;

$$x^3y''' = y^{\bullet\bullet\bullet} - 3y^{\bullet\bullet} + 2y^{\bullet}. \tag{3.80}$$

Substituting (3.79), and (3.78) back into (3.74), will be transformed into (3.75),

$$a_2y'' + (a_1 - a_2)y' + a_0y = g(x).$$
$$\text{Or, } a_2y^{\bullet\bullet} + (a_1 - a_2)y^{\bullet} + a_0y = g(x).$$

The Cauchy-Euler equation (3.75) can be extended to a third order equation as:

$$a_3y''' + a_2y'' + a_1y' + a_0y = g(x), \quad x > 0. \tag{3.81}$$

Which was formed by using (3.78), (3.79), and (3.80), and transformed into the following non-homogeneous equation with constant coefficient:

$$a_3y''' + (a_2 - 3a_3)y'' + (2a_3 - a_2 + a_1)y' + a_0y = g(x).$$
$$a_3y^{\bullet\bullet\bullet} + (a_2 - 3a_3)y^{\bullet\bullet} + (2a_3 - a_2 + a_1)y^{\bullet} + a_0y = g(x)$$

Example - 16:
Find the general solution for the Cauchy-Euler equation:
$$x^2 y'' + 6xy' + 4y = 0. \qquad (3.82)$$

Solution:
Since this is second order equation with non-constant (variable) coefficients, then using (3.78), and (3.79),will transform it into equation with constant coefficients:

$$y^{\bullet\bullet} - y^{\bullet} + 6y^{\bullet} + 4y = 0. \qquad (3.83)$$
$$y^{\bullet\bullet} + 5y^{\bullet} + 4y = 0. \qquad (3.84)$$

Equation (3.84) is an equation with constant coefficient, which has a solution of the form $y = e^{rt}$, differentiating twice and substituting back into (3.84) gives,

$$r^2 + 5r + 4 = 0. \qquad (3.85)$$

This is the auxiliary equation for the problem (3.82), using quadratic formula we can find the auxiliary roots: $r_1 = -4$, and $r_2 = -1$. Substituting we can write the general solution:

$$\begin{aligned} y(t) &= c_1 e^{r_1 t} + c_2 e^{r_2 t}. \\ &= c_1 (e^t)^{-4} + c_2 (e^t)^{-1}. \\ Then,\ y(t) &= c_1 x^{-4} + c_2 x^{-1}.\ x \neq 0. \end{aligned}$$

Where $e^t = x$.

Example - 17: Cauchy- Euler IVP
Solve the initial value problem:
$$x^2 y'' - 3xy' + 3y = 0. \qquad (3.86)$$

With initial conditions:
$$y(1) = -1,\ y'(1) = -9. \qquad (3.87)$$

Solution:
With the substitution of $x = e^t$ and its derivatives we get,
$$\ddot{y} - 4\dot{y} + 3y = 0.$$
And the auxiliary equation is:
$$r^2 - 4r + 3 = 0.$$
Which gives, by using the quadratic formula, the two auxiliary roots: $r_1 = 1$, and $r_2 = 3$. Then the general solution is:
$$y(t) = c_1 e^t + c_2 e^{3t}.$$
$$\text{And } y(x) = c_1 x + c_2 x^3.$$
To use the initial values, we differentiate the general solution once with respect to x.
$$y'(x) = c_1 + 3c_2 x^2.$$
Then applying the initial values, we get a system of two linear equations:
$$y(1) = -1 = c_1 + c_2.$$
$$y'(1) = -9 = c_1 + 3c_2.$$
Solving the linear system gives the values of $c_1 = 3$, and $c_2 = -4$. Then the general solution for the IVP is:
$$y(x) = 3x - 4x^2.$$

Example - 18:
Find the general solution for the third ODE Cauchy-Euler equation:
$$x^3 y'''' - 6x^2 y'' + 3xy' + 21y = 0. \qquad (3.88)$$

Solution:
Since this is a third order equation with non-constant (variable) coefficients, then it can be transformed into equation with constant coefficient as follows:

$$y^{\bullet\bullet\bullet} - 9y^{\bullet\bullet} + 11y^{\bullet} + 21y = 0. \qquad (3.89)$$

With auxiliary equation,

$$r^3 - 9r^2 + 11r + 21 = 0.$$

Here, using long division or synthetic division with the help of Descartes rule for the potential zero's :

```
-1 | 1   -9    11    21
         -1    10   -21
3  | 1  -10    21     0
          3   -21
7  | 1   -7     0
          7
     1    0
```

The division gives 3-zero's : $r_1 = -1$, $r_2 = 3$, and $r_3 = 7$. Then the general solution is:

$$y(t) = c_1 e^{-t} + c_2 e^{3t} + c_3 e^{7t}.$$
$$And,\ y(x) = c_1 x^{-1} + c_2 x^3 + c_3 x^7, for\ x \neq 0.$$

Using Cauchy Euler's Method to solve a Reduction of Order Problem

Example - 19: For the differential equation,

$$x^2 y'' + 6xy' + 6y = 0, \quad for \ x \geq 0. \qquad (3.90)$$

One solution is given $y_1 = x^{-2}$, Find the general solution for the problem.

A. Using Cauchy- Euler Method: Using the following substitutions:

$$x^2 y'' = y^{\bullet\bullet} - y^{\bullet}.$$
$$xy' = y^{\bullet}.$$

Equation(3.90) becomes,

$$y^{\bullet\bullet} - y^{\bullet} + 6y^{\bullet} + 6y = 0.$$
$$Or, \ y^{\bullet\bullet} + 5y^{\bullet} + 6y = 0.$$

And, using the substitution of $y = e^{rt}$ will lead to the auxiliary equation,

$$r^2 + 5r + 6 = 0.$$

With the following zero's: $r_1 = -3$, and $r_2 = -2$ from quadratic formula. Thus the general solution is:

$$y(t) = c_1 e^{-3t} + c_2 e^{-2t}.$$
$$Or, \ y(x) = c_1 x^{-3} + c_2 x^{-2}.$$

Where, $x = e^t$, c_1, and c_2 are constants, with $c_1 = 0$, and $c_2 = 1$ to make the second substitution be Linearly Independent.

B. Using Reduction of Order's Method:

Here we let the second solution be:

$$y_2 = uy_1 = ux^{-2} \tag{3.91}$$

Differentiation (3.91) twice and substituting back into equation (3.90) gives,

$$u'' + 2u'x^{-1} = 0. \tag{3.92}$$

Which upon making the following substitution,

$$u' = v.$$
$$\text{And, } u'' = v'.$$

Gives, the following equation,

$$v' + \frac{2}{x}v = 0. \tag{3.93}$$

equation (3.93) is a linear equation with $p(x) = \frac{2}{x}$, and integrating factor $I(x) = e^{2\int \frac{dx}{x}} = x^2$. Multiplying equation (3.93) by the integrating factor $I(x) = x^2$, and simplifying gives:

$$\int \frac{d}{dx}(x^2 v) = \int 0.$$
$$\text{Or, } v = cx^{-2} = u'.$$
$$\text{Then, } u = x^{-1}, \text{ assuming } c = 1.$$

The second solution is:

$$y_2 = uy_1.$$
$$= x^{-1}x^{-2}.$$
$$\text{Or, } y_2 = x^{-3}.$$

The general solution is:

$$y_G = c_1 y_1 + c_2 y_2.$$
$$y_G = c_1 x^{-2} + c_1 x^{-3}$$

Note: Equation (3.93) is also a separable equation, can be solved by separations as follows:

$$\frac{dv}{dx} = -\frac{2}{x}v.$$
$$Or, \quad \frac{dv}{v} = -\frac{2}{x}dx.$$
$$And\ by\ integration\ gives\ lnv = -2lnx + c.$$
$$Or, \quad v = kx^{-2}.$$

C. Using direct formula: by using the direct formula for the second solution y_2 gives:

$$y_2 = y_1 \int \frac{e^{-\int p(x)dx}}{(y_1)^2} dx.$$
$$= (x^{-2}) \int \frac{e^{-\int \frac{6}{x}dx}}{(x^{-2})^2} dx.$$
$$= (x^{-2}) \int \frac{x^{-6}}{(x^{-2})^2} dx.$$
$$= x^{-2} \int x^{-2} dx.$$
$$= -x^{-3}.$$

Ignoring the attached constant, the second equation will be:

$$y_2 = x^{-3}.$$

3.9 Technical Writing

Students will choose a topic from real life as an application for the second ODE with constant coefficients:

$$ay'' + by' + cy = g(t) \qquad (3.94)$$

Where, $a, b,$ and c are constants, and $g(t)$ is a given function depending on time. The best examples that applies to equation (3.94) are:
- The harmonic oscillation.
- The Electric Circuit.

Students will work on these two examples in two groups one for each. Each group will:
1. Write (derive) the DE for the model.
2. Write the general equation for the mechanical system.
3. Describe the motion of the system.
4. Give example to apply the steps.

3.10 Review Exercise

Wronskian:
1. Use the Wronskian to show that the following sets of functions are linearly independent:
a) $3, 2x, 3x^2$.
b) $sin2x, cosx$.
c) $2e^{2x}, e^x, xe^x$.
d) $e^{2x}, x+1$.
2. State that the set of functions are linearly dependent by stating their relation to each other: $sin2x, sinxcosx$.

Auxiliary Equations:
Write the auxiliary equation for the following problems (Do not solve):

3. $y'' - 2y' + 2y = 0.$
4. $z'' - 6z' = 0.$
5. $w''' - w'' + w' = 0.$

Reduction of Order:
6. Find the second linearly independent solution for the given differential equation:
$$y'' - y' - 2y = 0.$$
if the given solution is $y_1 = e^{-x}$.

7. Compute the second solution for the differential equation:
$$xy'' - (x+1)y' + y = 0, \quad x > 0 \text{ and } y_1 = e^x.$$

Homogeneous eq's with constant coefficients:
a) Real and distinct roots: Find the general solution for the following differential equations:

8. $y'' + 6y' - 7y = 0.$
9. $u'' - \frac{1}{4}u = 0.$
10. $y'' - \frac{5}{2} + \frac{3}{2}y = 0.$
11. $y''' + 3y'' - y' - 3 = 0.$
12. $y''' + y'' - 6y' = 0.$

b) Repeated roots: Find the general solution for the following differential equations:

13. $y'' + y' + \frac{1}{4}y = 0.$
14. $y'' - 2y + y = 0.$
15. $y'' + 6y' + 9y = 0.$

3.10. REVIEW EXERCISE

c) Complex roots: Find the general solution for the following differential equations:

$$16. \quad y^{(4)} + 5y'' + 4y = 0.$$
$$17. \quad u'' + 9u = 0.$$
$$18. \quad y'' + 4y' + 6y = 0.$$
$$19. \quad w'' - 5w' + 7w = 0.$$

Initial Value Problem:
20. Solve the IVP :

$$u'' + ku' - 5u = 0.$$

With initial values: $u(0) = 1$, $u'(0) = 0$, and $k = -1, 2, 3, 6$.

Un-determined coefficients Method
Superposition
Solve the non-homogeneous problems:

$$21. \quad w'' - w' + w = \sin t.$$
$$22. \quad y'' - y' + 2y = 2x.$$
$$23. \quad u'' - u = 3x^2.$$

Variation of Parameters:
24. Find the general solution for the problem :

$$u'' + u = \tan t, \text{ on the interval } I = \{-\pi/2, \pi/2\}$$

Cauchy-Euler Method:
25. Solve the differential equation :

$$u'' - \frac{3}{x}u' + \frac{6}{x^2}u = 0.$$

3.11 Review Exercise Solutions

Wronskian:

1.

a) $W(3, 2x, 3x^2) = \begin{bmatrix} 3 & 2x & 3x^2 \\ 0 & 2 & 6x \\ 0 & 0 & 6 \end{bmatrix}$
$= 36 \neq 0.$

b) $W(sin2x, cosx) = \begin{bmatrix} sin2x & cosx \\ 2cos2x & -sinx \end{bmatrix}$
$= -2cos^3 x \neq 0.$

c) $W(2e^{2x}, e^x, xe^x) = \begin{bmatrix} 2e^{2x} & e^x & xe^x \\ 4e^{2x} & e^x & xe^x + e^x \\ 8e^{2x} & e^x & xe^x + 2e^x \end{bmatrix}$
$= 2e^{4x} \neq 0.$

d) $W(e^{2x}, x+1) = \begin{bmatrix} e^{2x} & x+1 \\ 2e^{2x} & 1 \end{bmatrix}$
$= -e^{2x}(1-x) \neq 0.$

2. Since $sin2x = 2sinx\ cosx$, then this means that the first function $sin2x$ is twice the second one. also we can prove that their Wronskian is equal zero.

$W(sin2x, sinx\ cosx) = \begin{bmatrix} sin2x & sinx\ cosx \\ 2cos2x & cos^2 x - sin^2 x \end{bmatrix}$
$= 0.$

3.11. REVIEW EXERCISE SOLUTIONS

Auxiliary Equations:

$$3. \quad r^2 - 2r + 2 = 0.$$
$$4. \quad r^2 - 6r = 0.$$
$$5. \quad r^3 - r^2 + r + 3 = 0.$$

Reduction of Order:

6. a) Using the formula directly gives,

$$y_2 = y_1 \int \frac{e^{-\int p(x)dx}}{(y_1)^2} \, dx.$$

$$= e^{-x} \int \frac{e^{-\int -1 dx}}{(e^{-x})^2} \, dx.$$

$$= e^{-x} \int e^{3x} \, dx.$$

$$= e^{-x} \{\frac{1}{3} e^{3x}\}.$$

$$= \frac{1}{3} e^{2x}.$$

Neglecting the constant we get the second solution $y_2 = e^{2x}$ And the general solution is:

$$y_G = c_1 y_1 + c_2 y_2.$$
$$y_g = c_1 e^{-x} + c_2 e^{2x}.$$

b) The equation can also be solved by using the auxiliary equation:

$$r^2 - 2r + 2 = 0.$$
$$(r - 2)(r + 1) = 0.$$

Then, $r_1 = -1,$ and $r_2 = 2.$

And, the general solution is $y_G = c_1 e^{-x} + c_2 e^{2x}.$

7. After rewriting the given equation in the standard form by dividing by x,
$$y'' - \frac{x+1}{x}y' + \frac{1}{x}y = 0.$$

we can use the formula directly to compute the second solution as:

$$y_2 = y_1 \int \frac{e^{-\int p(x)dx}}{(y_1)^2} \, dx.$$

$$= e^x \int \frac{e^{\int \frac{x+1}{x}}}{(e^x)^2} \, dx.$$

Then the integrating by parts gives : $y_2 = x + 1$.

Homogeneous equations with constant coefficients:

a. Real and distinct roots:

8. The auxiliary equation for the problem is:
$$r^2 + 6r - 7 = 0.$$
$$with \ factors : \ (r+7)(r-1) = 0.$$
$$And \ roots : \ r_1 = -7, \ r_2 = 1.$$
Then the general solution
is : $y_G = c_1 e^{-7x} + c_2 e^x$.

9. The auxiliary equation for the problem is:
$$4r^2 - 1 = 0.$$
$$with \ factors : \ (2r+1)(2r-1) = 0.$$
$$And \ roots : \ r_1 = -1/2, \ r_2 = 1/2.$$
Then the general solution is :
$$y_G = c_1 e^{-x/2} + c_2 e^{x/2}.$$

3.11. REVIEW EXERCISE SOLUTIONS

10. The auxiliary equation for the problem is:

$$2r^2 - 5r + 3 = 0.$$
$$with\ factors:\ (2r-3)(r-1) = 0.$$
$$And\ roots:\ r_1 = 3/2,\ r_2 = 1.$$

Then the general solution is : $y_G = c_1 e^{3x/2} + c_2 e^x$.

11. The auxiliary equation for the problem is:

$r^3 + 3r^2 - r - 3 = 0.$
with factors: $(r+3)(r-1)(r+1) = 0.$
And roots: $r_1 = -3,\ r_2 = 1,\ r_3 = -1.$
Then the general solution is:
$y_G = c_1 e^{-3x} + c_2 e^x + c_3 e^{-x}.$

12. The auxiliary equation for the problem is:

$r^3 + r^2 - 6r = 0.$
with factors: $r(r+3)(r-2) = 0.$
And roots: $r_1 = 0, r_2 = -3, r_3 = 2.$
Then the general solution is:
$y_G = c_1 + c_2 e^{-3x} + c_3 e^{2x}.$

b. Repeated roots:

13. The auxiliary equation for the problem is:

$$r^2 + r + 1/4 = 0.$$
$$with\ factors:\ (r+1/2)(r+1/2) = 0.$$
$$And\ roots:\ r_1 = -1/2,\ r_2 = -1/2.$$

Then the general solution is : $y_G = c_1 e^{-x/2} + c_2 x e^{-x/2}$.

14. The auxiliary equation for the problem is:

$$r^2 - 2r + 1 = 0.$$

$$\text{with factors}: (r-1)(r-1) = 0.$$
$$\text{And roots}: r_1 = 1, \ r_2 = 1.$$
$$\text{Then the general solution is}: y_G = c_1 e^x + c_2 x e^x.$$

15. The auxiliary equation for the problem is:

$$r^2 + 6r + 9 = 0.$$
$$\text{with factors}: (r+3)(r+3) = 0.$$
$$\text{And roots}: r_1 = -3, \ r_2 = -3.$$
$$\text{Then the general solution is}: y_G = c_1 e^{-3x} + c_2 x e^{-3x}.$$

c. Complex roots:

16. The auxiliary equation for the problem is:

$$r^4 + 5r^2 + 4 = 0.$$

To simplify more, we let $r^2 = u$, then $r^4 = u^2$ then the equation in terms of u is:

$$u^2 + 5u + 4 = 0.$$
$$\text{with factors}: (u+4)(u+1) = 0.$$
$$\text{And roots}: u_1 = -4 = 1^2, \ u_2 = -1 = r_2^2.$$
$$\text{And the roots } r_{1,2} = \pm 2i, \text{ and } r_{3,4} = \pm i.$$
$$\text{Then the general solution is}:$$
$$y_G = c_1 e^{2ix} + c_2 e^{-2ix} + c_3 e^{-ix} + c_4 e^{1x}.$$
$$\text{Or, } y_G = c_1 \cos 2x + c_2 \sin 2x + c_3 \cos x + c_4 \sin x.$$

17. The auxiliary equation for the problem is:

$r^2 + 9 = 0.$
$r^2 = -9.$
Then, $r = \mp 3i.$

3.11. REVIEW EXERCISE SOLUTIONS

And roots: $r_1 = -3i, r_2 = 3i$.
Then the general solution is:
$y_G = c_1 e^{3ix} + c_2 e^{-3ix}$.
Or, $y_G = c_1 cos 3x + c_2 sin 3x$.

18. The auxiliary equation for the problem is:
$$r^2 + 4r + 6 = 0.$$
$Using \ Quadratic \ formula \ gives: \quad r = -2 \pm \sqrt{2}i.$
$Then \ the \ general \ solution \ is:$
$$y_G = c_1 e^{(-2+\sqrt{2}i)x} + c_2 e^{(-2-\sqrt{2}i)x}.$$
$Or, \ y_G = e^{-2x}\{c_1 cos\sqrt{2}x + c_2 sin\sqrt{2}x\}.$

19. The auxiliary equation for the problem is:
$$r^2 - 5r + 7 = 0.$$
$Using \ Quadratic \ formula \ gives: \quad r = \dfrac{5}{2} \pm \dfrac{\sqrt{3}}{2}i.$
$Then \ the \ general \ solution \ is:$
$$y_G = c_1 e^{(\frac{5}{2}+\frac{\sqrt{3}}{2}i)x} + c_2 e^{(\frac{5}{2}-\frac{\sqrt{3}}{2}i)x}.$$
$Or, \ y_G = e^{\frac{5}{2}}\{c_1 cos\sqrt{3}x + c_2 sin\sqrt{3}x\}.$

Initial Value Problem:

20. Solve the IVP :
$$u'' + ku' - 5u = 0.$$
With initial values: $u(0) = 1$, $u'(0) = 0$, and $k = -1, 2, 3, 6$. When $k = -1$, the equation is:
$$u'' - u' - 5u = 0.$$

And its auxiliary equation is:

$$r^2 - r - 5 = 0.$$

With roots: $r = \frac{1\pm\sqrt{21}}{2}$. And general solution :

$$y_G = c_1 e^{\frac{1+\sqrt{21}}{2}x} + c_2 e^{\frac{1-\sqrt{21}}{2}x}.$$

Differentiating the general solution gives:

$$y'_G = (\frac{1+\sqrt{21}}{2})c_1 e^{\frac{1+\sqrt{21}}{2}x} + (\frac{1-\sqrt{21}}{2})c_2 e^{\frac{1-\sqrt{21}}{2}x}.$$

Applying the given initial values to y_G, and y'_G form a system of two equations, which upon solving leads to the value of the constants c_1, and c_2 as follows:

$$1 = c_1 + c_2.$$
$$0 = \frac{1+\sqrt{21}}{2}c_1 + \frac{1-\sqrt{21}}{2}c_2.$$

Then $c_1 = \frac{\sqrt{21}-1}{2\sqrt{21}}$, and $c_2 = \frac{1+\sqrt{21}}{2\sqrt{21}}$.

Thus the general solution for the IVP is:

$$y_G = \frac{\sqrt{21}-1}{2\sqrt{21}} e^{\frac{1+\sqrt{21}}{2}x} + \frac{1+\sqrt{21}}{2\sqrt{21}} e^{\frac{1-\sqrt{21}}{2}x}.$$

When $k = 2, 3, 6$ is left for the students to solve.

Un-determined coefficients Method:

21. The general solution for this type of problem is :

3.11. REVIEW EXERCISE SOLUTIONS

$y_G = y_H + y_p$. Where, $y_H \Rightarrow$ is the solution of:
$w'' - w' + w = 0$, with auxiliary equation:

$$r^2 - r + 1 = 0.$$

Using quadratic formula gives : $r = \dfrac{1}{2} \pm \dfrac{\sqrt{3}}{2}i.$

Then, $w_H = e^{t/2}\{c_1\cos\dfrac{\sqrt{3}}{2}t + c_2\sin\dfrac{\sqrt{3}}{2}t\}.$

To find w_p we let (see the table for y_p choices),

$$\begin{aligned} w_p &= A\sin t + B\cos t. \\ w'_p &= A\cos t - B\sin t. \\ w''_p &= -A\sin t - B\cos t. \end{aligned}$$

Substituting all back into the original non-homogeneous equation: $w'' - w' + w = \sin t$, and solving for the constants A, and B, we find that $A = 0$, and $B = 1$. Then $w_p = \cos t$.
And the general solution for the problem is:

$$\begin{aligned} w_G &= w_H + w_p. \\ w_G &= e^{t/2}\{c_1\cos\dfrac{\sqrt{3}}{2}t + c_2\sin\dfrac{\sqrt{3}}{2}t\} + \cos t. \end{aligned}$$

22. The general solution for this type of problem is : $y_G = y_H + y_p$. Where, $y_H \Rightarrow$ is the solution of $y'' - y' + 2y = 0$, with auxiliary equation:

$$r^2 - r + 2 = 0.$$

Using the quadratic formula gives : $r = \dfrac{1}{2} \pm \dfrac{\sqrt{7}}{2}i.$

Then, $y_H = e^{x/2}\{c_1\cos\dfrac{\sqrt{7}}{2}x + c_2\sin\dfrac{\sqrt{7}}{2}x\}.$

To find y_p we let (see the table for y_p choices):

$$y_p = Ax + B$$
$$y'_p = A.$$
$$y''_p = 0.$$

Substituting all back into the original non-homogeneous equation: $y'' - y' + 2y = 2x$, and solve for the constants A, and B, we find that $A = 1$, and $B = \frac{1}{2}$. Then, $y_p = x + \frac{1}{2}$. And, the general solution for the problem is:

$$y_G = y_H + y_p.$$
$$y_G = e^{x/2}\{c_1 \cos\frac{\sqrt{7}}{2}x + c_2 \sin\frac{\sqrt{7}}{2}x\} + x + \frac{1}{2}.$$

23. The general solution for this type of problem is : $u_G = u_H + u_p$. Where, $u_H \Rightarrow$ is the solution of $u'' - u = 0$, with auxiliary equation:

$$r^2 - 1 = 0. \text{ with roots } r = \pm 1.$$
$$\text{Then,} \quad u_H = c_1 e^{-x} + c_2 e^x.$$

To find u_p we let (see the table for y_p choices):

$$u_p = Ax^2 + Bx + C$$
$$u'_p = 2Ax + B.$$
$$u''_p = 2A.$$

Substituting all back into the original non-homogeneous equation, we find that $A = -3, B = 0, C = -6$. Thus, $u_p = -3x^2 - 6$.
Then the general solution for the problem is:

$$u_G = u_H + u_p.$$
$$u_G = c_1 e^{-x} + c_2 e^x - 3x^2 - 6.$$

3.11. REVIEW EXERCISE SOLUTIONS

Variation of Parameters:

24. The general solution for this problem is : $u_G = u_H + u_p$. Where, $u_H \Rightarrow$ is the solution of $u'' + u = 0$, with auxiliary equation:

$$r^2 + 1 = 0. \text{ with roots } r = \pm i.$$
$$\text{Then,} \quad u_H = c_1 cost + c_2 sint.$$

Then: $u_1 = cost$, and $u_2 = sint$. To find u_p by using the variation of parameter method, we let:

$$u_p = vu_1 + wu_2.$$

Then differentiating twice and substituting back into the original non-homogeneous DE, leads to a system of two equations:

$$v'u_1 + w'u_2 = 0.$$
$$v'u_1' + w'u_2' = g(t).$$

Using Cramer's we can solve for the unknowns v, and w, where $u_1 = cost$, and $u_2 = sint$ from u_H, and $g(t) = tant$. From the given differential equation;

$$D = \begin{bmatrix} u_1 & u_2 \\ u_1' & u_2' \end{bmatrix}$$
$$= \begin{bmatrix} cost & sint \\ -sint & cost \end{bmatrix}$$
$$= 1 \neq 0.$$

$$\text{Then, } v' = \begin{bmatrix} 0 & sint \\ tant & cost \end{bmatrix}$$

$$= -sint\, tant.$$
$$v = \int \frac{cos^2 t - 1}{cost} dt.$$
$$= \int cost\, dt - \int sect\, dt.$$
$$\text{Or, } v = sint - ln\,|\,sect + tant\,|.$$

In similar way:
$$w' = \begin{bmatrix} cost & 0 \\ -sint & tant \end{bmatrix}$$
$$= cost\, tant = cost\frac{sint}{cost}.$$
$$\text{Then, } w = \int sint\, dt.$$
$$\text{Or, } w = -cost.$$

Then the particular solution for the problem is:
$$u_p = vu_1 + wu_2.$$
$$= cost\{sint - ln\,|\,sect + tant\,|\}$$
$$+ sint\{-cost\}.$$
$$u_p = cosx\{ln\,|\,sect + tant\,|\}.$$

Thus, the general solution is:
$$u_G = u_H + u_p.$$
$$u_G = c_1 cost + c_2 sint + cost\{ln\,|\,sect + tant\,|\}.$$

Cauchy-Euler Method:

25. Rewriting the equation in Cauchy-Euler form as:
$$x^2 u'' - 3xu' + 6u = 0.$$

Then applying the substitution $x = e^t$, with chain rule to the standard form equation gives:

$$x^2 u'' = y^{\bullet\bullet} - y^{\bullet}, \text{ and } xu' = y^{\bullet}.$$

And substituting back gives,

$$t^2 u'' - 3tu' + 6u = 0.$$
$$u^{\bullet\bullet} - u^{\bullet} - 3u^{\bullet} + 6u = 0.$$
$$\text{Or } u^{\bullet\bullet} - 4u^{\bullet} + 6u = 0.$$

Then the auxiliary equation is, $r^2 - 4r + 6 = 0$. Where, $r = 2 \pm \sqrt{2}i$. And the general solution is,

$$u(t)_G = e^{2t}\{c_1 \cos\sqrt{2}t + c_2 \sin\sqrt{2}t\}.$$

But $x = e^t$, then,

$$u(x)_G = x^2\{c_1\cos(\sqrt{2}lnx) + c_2\sin\{\sqrt{2}lnx\}.$$

3.12 Chapter-3 Assignment

1. Use the Wronskian to show if the following sets of functions are linearly dependent or independent:

a) $\quad y_1 = x^2\cos(lnx).$
$\quad\quad y_2 = x^2\sin(lnx).$

b) $\quad y_1 = \tan^2 x - \sec^2 x.$
$\quad\quad y_2 = 3.$

2. Using the reduction of order method, find a second solution for the given differential equation:

$$xy'' + (1 - 2x)y' + (x - 1)y = 0. \; x > 0.$$
$$\text{If the given solution is, } y_1 = e^x.$$

3. Solve the homogeneous equation of constant coefficients:

$$4w'' + 20w' + 25w = 0.$$

4. Using Cauchy-Euler Method find the general solution for the following problems:

$$a) \quad x^2 y'' + 3xy' + 5y = 0.$$
$$b) \quad x^3 y''' + x^2 y'' + 5xy' - 27y = 0.$$

5. Using the superposition method for the undetermined coefficient, find the general solution for:

$$y'' - y' + y = \sin x - 3e^{2x}.$$

6. Solve the initial value problem:

$$y'' - 7y' + 10y = x^2 - 4 + e^x.$$
$$With \ y(0) = 3, and \ y'(0) = -3.$$

Chapter 4

The Laplace Transform

4.1 Introduction

Pierre Simon de Laplace (1749 - 1827),a French mathematician was a theoretical Astronomer too. The fundamental equation of theory was named after him "The Laplacian equation",

$$\frac{\partial^2 u}{\partial x^2} + \frac{\partial^2 u}{\partial y^2} + \frac{\partial^2 u}{\partial z^2} = 0. \quad (4.1)$$

Using the Laplacian operator symboled as "$\triangledown$", where

$$\triangledown = \frac{\partial^2}{\partial x^2} + \frac{\partial^2}{\partial y^2} + \frac{\partial^2}{\partial z^2}. \quad (4.2)$$

Then, equation (4.1) can be written as:

$$\triangledown u = 0. \quad (4.3)$$

Use of Laplace Transform: In mathematics, when dealing with difficult problems, it is useful to transform the problem so as to obtain a simpler problem that can be solved, then the solution of the simpler problem can be transformed to find the solution of the original problem as seen in solving Cauchy-Euler Problems.

Up to this chapter we have dealt with continuous functions, but dealing with discontinuous functions or periodic can be complicated, situation like this arises in practical problems such as electrical engineering, but Laplace transform acts as a powerful tool to solve such problems. In this chapter we will explore the transformation of an initial-value problem into an Algebraic equation, which can be solved easily, and then using the inverse Laplace will transform it back to the original problem.

Laplace Transform Symbol: Throughout this chapter, Our symbol for Laplace transform will be: "$\mathcal{L}$".

4.2 Definition of Laplace Transforms

Laplace transform takes any function of t, and produces a function of s, and this is done by multiplying $f(t)$ by e^{-st} and integrating with respect to t from 0 to ∞ as written below:

$$\mathcal{L}\{f(t)\} = \int_0^\infty f(t)e^{-st}dt = F(s). \qquad (4.4)$$

The Laplace transform of basic functions is a routine computation done in the same way as the integration of basic functions such as Polynomials, Rational, and

4.2. DEFINITION OF LAPLACE TRANSFORMS

Trigonometric functions, and are provided in tables, people use these tables directly to get transformation of any type of functions. We advise students to practice on using few to learn the method before they start using tables directly. Here we will present the transformation of basic functions:

Example-1: Laplace transform of $f(t) = a$.
Here a is a constant. Then using the formula from (4.4) as:

$$\begin{aligned}
\mathcal{L}\{a\} &= \int_0^\infty a e^{-st} dt. \\
&= a \int_0^\infty e^{-st} dt. \\
&= a \lim_{b \to \infty} \{-\frac{1}{s} e^{-st} |_0^b\}. \\
&= a\{-\frac{1}{s}(0-1)\} = \frac{a}{s}.
\end{aligned}$$

Then, $\mathcal{L}\{a\} = \frac{a}{s}$.

Example-2: Laplace transform of $f(t) = 2$.

$$\begin{aligned}
\mathcal{L}\{f(t)\} = \mathcal{L}\{2\} &= \int_0^\infty 2 e^{-st} dt. \\
&= 2 \int_0^\infty e^{-st} dt. \\
&= 2 \lim_{b \to \infty} \{-\frac{1}{s} e^{-st} |_0^b\}. \\
&= 2\{-\frac{1}{s}(0-1)\} = \frac{2}{s}.
\end{aligned}$$

Then, $\mathcal{L}\{2\} = \frac{2}{s}$.

Example-3: Laplace transform of $f(t) = e^{at}$.
Applying (4.4) gives:

$$\mathcal{L}\{e^{at}\} = \int_0^\infty e^{at} e^{-st} dt.$$

$$= \int_0^\infty e^{(a-s)t} dt.$$

$$= \lim_{b \to \infty} \{-\frac{1}{a-s} e^{(a-s)t} \big|_0^b\}.$$

$$= a\{-\frac{1}{s}(0-1)\} = \frac{1}{s-a}. \quad s > a.$$

Then, $\mathcal{L}\{e^{at}\} = \dfrac{1}{s-a}$.

Example-4: Laplace transform of $f(t) = e^{2t}$.

$$\mathcal{L}\{e^{2t}\} = \int_0^\infty e^{2t} e^{-st} dt.$$

$$= \int_0^\infty e^{(2-s)t} dt.$$

$$= \lim_{b \to \infty} \{-\frac{1}{2-s} e^{(2-s)t} \big|_0^b\}.$$

$$= \{-\frac{1}{2-s}(0-1)\} = \frac{1}{s-2}. \quad s > 2.$$

Then, $\mathcal{L}\{e^{2t}\} = \dfrac{1}{s-2}$.

Example-5: Laplace transform of $f(t) = \sin at$.
Applying (4.4) gives:

$$\mathcal{L}\{\sin at\} = \int_0^\infty \sin at \, e^{-st} dt.$$

$$\mathcal{L}\{\sin at\} = \frac{a}{s^2 + a^2}.$$

Example-6: Laplace transform of $f(t) = \sin 4t$.
Applying (4.4) gives:

$$\mathcal{L}\{\sin 4t\} = \int_0^\infty \sin 4t \, e^{-st} dt.$$

4.2. DEFINITION OF LAPLACE TRANSFORMS

$$\mathcal{L}\{sin\ 4t\} = \frac{4}{s^2 + 4^2}.$$

$$\mathcal{L}\{sin\ 4t\} = \frac{4}{s^2 + 16}.$$

Example-7: Laplace transform of $f(t) = 5 + 1/2e^{2t}$.

$$\begin{aligned}
\mathcal{L}\{f(t)\} &= \mathcal{L}\{5\} + \mathcal{L}\{1/2e^{2t}\}. \\
&= \int_0^\infty 5e^{-st}\,dt + \int_0^\infty \frac{1}{2}e^{2t}e^{-st}\,dt. \\
&= 5\int_0^\infty e^{-st}\,dt + \frac{1}{2}\int_0^\infty e^{(2-s)t}\,dt. \\
&= 5\lim_{b\to\infty}\{-\frac{1}{s}e^{-st}\ |_0^b\} \\
&+ \frac{1}{2}\lim_{b\to\infty}\{-\frac{1}{2-s}e^{(2-s)t}\ |_0^b\}. \\
&= 5\{-\frac{1}{s}(0-1)\} + \frac{1}{2}\{-\frac{1}{2-s}(0-1)\}.
\end{aligned}$$

$$Then,\ \mathcal{L}\{f(t)\} = \frac{5}{s} + \frac{1/2}{s-2}.$$

In the same manner the student can find the rest of the transformed functions listed on the following table for $n > 0, a > 0$:

f(t)	F(s)	f(t)	F(s)
a	$\frac{a}{s}$	t^n	$\frac{n!}{s^{n+1}}$
e^{at}	$\frac{1}{s-a}$	$e^{at}t^n$	$\frac{n!}{(s-a)^{n+1}}$
$sin(at)$	$\frac{a}{s^2+a^2}$	$cos(at)$	$\frac{s}{s^2+a^2}$
$e^{at}sin(bt)$	$\frac{b}{(s-a)^2+b^2}$	$e^{at}cos(bt)$	$\frac{s-a}{(s-a)^2+b^2}$

Table(2)

4.3 Properties of Laplace Transform

1. Laplace transform is a linear operator:
$\mathcal{L}\{f(t) + g(t)\} = \mathcal{L}\{f(t)\} + \mathcal{L}\{g(t)\}$

2. $\mathcal{L}\{c\, f(t)\} = c\, \mathcal{L}\{f(t)\} = c\, F(s)$.

3. If , $\mathcal{L}\{f(t)\} = F(s)$, then: $\mathcal{L}\{t f(t)\} = -\frac{d}{ds} F(s)$.

4. For each positive integer n: $\mathcal{L}\{t^n f(t)\} = (-1)^n \frac{d^n F}{ds^n}(s)$.

5. For a continuous function $f(t)$: $\mathcal{L}\{f'(t)\} = s\mathcal{L}\{f(t)\} - f(0)$.

6. If $f(t)$, and $f'(t)$ are both continuous then:
$\mathcal{L}\{f''(t)\} = s^2 \mathcal{L}\{f(t)\} - sf(0) - f'(0)$.

7. If $f(t)$, and $f'(t)$ and $f''(t)$ are all continuous then:
$\mathcal{L}\{f'''(t)\} = s^3 \mathcal{L}\{f(t)\} - s^2 f(0) - sf'(0) - f''(0)$.
Or in general for n-terms:

8. $\mathcal{L}\{f^n(t)\} = s^n \mathcal{L}\{f(t)\} - s^{n-1} f(0) - s^{n-2} f'(0) ... - f^{n-1}(0)$.

9. Let $\mathcal{L}\{f(t)\} = F(s)$, then: $\mathcal{L}\{\int_0^t f(x)dx\} = \frac{1}{s} F(s)$.

10. Let $\mathcal{L}\{f(t)\} = F(s)$, then: $\mathcal{L}\{e^{at} f(t)\} = F(s-a)$.

11. $\mathcal{L}\{f(t) u(t-a)\} = e^{-as} \mathcal{L}\{f(t+a)\}$.

12. $\mathcal{L}\{t^n f(t)\} = (-1)^n \frac{d^n F}{ds^n}$.

13. $\lim_{s \to \infty} \mathcal{L}\{f\}(s) = 0$.

14. $\mathcal{L}\{e^{at} f(t)\} = F(s-a)$.

15. $\mathcal{L}^{-\infty}\{\frac{d^n F}{ds^n}\} = (-t)^n f(t)$.

Table(3), Some Laplace Transform Properties.

4.4 Inverse Laplace Transform

we have seen in section (4.2) how the Laplace transform as an integral operator transforms a function $f(t)$ into a function $F(s)$. Now in this section we will try to restore the original function $f(t)$ from the transformed function $F(s)$ by using the **Inverse Laplace Transform**. it is clear that the operator that yields $f(t)$ from $F(s)$ is called the **Inverse Laplace Transform** which is denoted by $\mathcal{L}^{-1}$. This can be written symbolically as:

$$f(t) = \mathcal{L}^{-1}[F(s)].$$
$$Alternatively,\ f(t) = \mathcal{L}^{-1}[L\{f(t)\}].$$

The following table shows some basic transformed functions $F(s)$ being transformed back into the original functions $f(t)$.

$f(t)$	$F(s)$	$f(t)$	$F(s)$
$\frac{a}{s}$	a	$\frac{n!}{s^{n+1}}$	$t^n,\ n = 0, 1, \ldots$
$\frac{1}{s-a}$	e^{at}	$\frac{n!}{(s-a)^{n+1}}$	$e^{at}t^n$
$\frac{a}{s^2+a^2}$	$sin(at)$	$\frac{s}{s^2+a^2}$	$cos(at)$
$\frac{b}{(s-a)^2+b^2}$	$e^{at}sin(bt)$	$\frac{s-a}{(s-a)^2+b^2}$	$e^{at}cos(bt)$

Table(4)

<u>Now</u>, we are ready to apply the Laplace Transform to linear differential equations with constant coefficients that is widely used in Electrical Engineering.

4.5 Solving Initial Value Problems

To solve IVP of linear differential equation, homogeneous or non homogeneous, the following steps are applied:

Step-1: $\mathcal{L}\{y(t)\} \to Y(s)$

In this step the DE is transformed from $y(t)$ by Laplace transform into an algebraic equation in s, then the algebraic equation is solved for the unknown $Y(s)$ by algebraic manipulations.

Step-2: $\mathcal{L}^{-1}\{Y(s)\} \to y(t)$.

In this step the unknown $Y(s)$ is transformed by the Inverse Laplace transform, and the result $y(t)$ is the solution to the original DE.

Example-1:
Consider the initial value problem:

$$y' - 4y = 3. \text{ with } y(0) = 2.$$

Applying the Laplace operator to the equation,

$$\mathcal{L}\{y' - 4y = 3\}.$$

Then distributing the operator, using $Y(s) = \mathcal{L}\{y(t)\}$, and $\mathcal{L}\{y'(t)\} = sY(s) - f(0)$. gives:

$$\mathcal{L}\{y'\} - 4\mathcal{L}\{y\} = \mathcal{L}\{3\}.$$
$$sY(s) - y(0) - 4Y(s) = \frac{3}{s}.$$
$$sY(s) - 2 - 4Y(s) = \frac{3}{s}.$$

4.5. SOLVING INITIAL VALUE PROBLEMS

$$(s-4)Y(s) = \frac{3}{s} + 2.$$

$$\text{Then, } Y(s) = \frac{2s+3}{s(s-4)}.$$

To restore $y(t)$, we have to apply the Inverse Laplace operator:

$$\mathcal{L}^{-1}[Y(s)] = \mathcal{L}^{-1}\{\frac{2s+3}{s(s-4)}\}.$$

Where $\{\frac{2s+3}{s(s-4)}\}$ is a rational that needs to be simplified first before applying the integral, by using the partial fraction method.

$$\frac{2s+3}{s(s-4)} = \frac{2s}{s(s-4)} + \frac{3}{s(s-4)}.$$

$$= \underbrace{\frac{A}{s} + \frac{B}{s-4}}_{1st\ fraction} + \underbrace{\frac{C}{s} + \frac{D}{s-4}}_{2nd\ fraction}.$$

Solving the fractions separately for A, B, C, and D gives: $A = 0, B = 2, C = -4/3, and\ D = 4/3$. Then,

$$\frac{2s+3}{s(s-4)} = \frac{A}{s} + \frac{B}{s-4} + \frac{C}{s} + \frac{D}{s-4}.$$

$$= \frac{2}{s-4} - \frac{4/3}{s} + \frac{4/3}{s-4}$$

Now applying the inverse Laplace,

$$\mathcal{L}^{-1}[Y(s)] = \mathcal{L}^{-1}\{\frac{2s+3}{s(s-4)}\}.$$

$$= \mathcal{L}^{-1}\{\frac{2}{s-4} - \frac{4/3}{s} + \frac{4/3}{s-4}\}.$$

$$\mathcal{L}^{-1}[Y(s)] = \mathcal{L}^{-1}\{\frac{2}{s-4}\} - \mathcal{L}^{-1}\{\frac{4/3}{s}\}$$
$$+ \mathcal{L}^{-1}\{\frac{4/3}{s-4}\}.$$
$$y(t) = 2e^{4t} - \frac{4}{3} + \frac{4}{3}e^{4t}.$$
$$y(t) = \frac{10}{3}e^{4t} - \frac{4}{3}.$$

Example-2:
Solve the initial value problem using Laplace Transform:
$$y'' + 4y = 0, \; with \; y(0) = A, \; and \; y'(0) = B.$$

Applying Laplace Transform gives:
$$\mathcal{L}\{y'' + 4y\} = 0\}.$$
$$\mathcal{L}\{y''\} + 4\mathcal{L}\{y\} = 0.$$
$$s^2 Y(s) - sy(0) - y'(0) + 4Y(s) = 0.$$
$$s^2 Y - sA - B + 4Y = 0.$$
$$Y(s^2 + 4) = As + B.$$
$$Then, \; Y(s) = \frac{As + B}{s^2 + 4}.$$

Now, applying the inverse Laplace to $Y(s)$ gives $y(t)$:
$$\mathcal{L}^{-1}\{Y(s)\} = \mathcal{L}^{-1}\{\frac{As+B}{s^2+4}\}.$$
$$= A\mathcal{L}^{-1}\{\frac{s}{s^2+4}\} + B\mathcal{L}^{-1}\{\frac{1}{s^2+4}\}.$$
$$= A\mathcal{L}^{-1}\{\frac{s}{s^2+4}\} + \frac{1}{2}B\mathcal{L}^{-1}\{\frac{2}{s^2+4}\}.$$
$$Then, y(t) = A\cos(2t) + \frac{1}{2}B\sin(2t).$$

Example-3:
Consider the initial value problem:
$$y' = e^t. \text{ with } y(0) = -1.$$

Applying Laplace Transform gives:
$$\mathcal{L}\{y'\} = e^t\}.$$
$$\mathcal{L}\{y'\} = \mathcal{L}\{e^t\}.$$
$$sY(s) - y(0) = \frac{1}{s-1}.$$
$$sY + 1 = \frac{1}{s-1}.$$
$$sY = \frac{1}{s-1} - 1.$$
$$sY = \frac{2-s}{s(s-1)}.$$
$$Y(s) = \frac{2}{s(s-1)} - \frac{1}{s-1}.$$

Applying the inverse Laplace leads to the solution:
$$y(t) = -2 + e^t.$$

4.6 Decomposition of Partial Fraction:

Notice from the previous examples that the inverse Laplace is applied to a rational function in s, which in some cases is complicated, and requires to be decomposed using calculus method called "Partial Fractions". Now we will review this method briefly. Recall from calculus the "Rational function":

$$R(s) = \frac{p(s)}{q(s)}, \text{ where } q(s) \neq 0.$$

In solving partial fractions there are three cases of $q(s)$ to be considered:

1. Non-repeated (distinct) linear factors:
The rational appears as:

$$R(s) = \frac{p(s)}{(s-r_1)(s-r_2)...(s-r_n)}.$$

Where $r_1, r_2, ...r_n$ are the distinct factors. Then to use the inverse Laplace the partial fraction will be decomposed as:

$$\frac{A_1}{s-r_1} + \frac{A_2}{x-r_2} + ... + \frac{A_n}{x-r_n}.$$

Example: Write the partial fraction decomposition of :

$$\frac{s}{s^2 - 2s - 8}$$

Solution: First we factor the denominator or $q(s)$:

$$q(s) = (s+2)(s-4).$$

Then decompose the rational expression:

$$\frac{s}{s^2 - 2s - 8} = \frac{A}{s+2} + \frac{B}{s-4}.$$

Where A, and B are constants to be determined. Solving the two fractions by taking the common denominator and write:

$$s = A(s-4) + B(s+2).$$

4.6. DECOMPOSITION OF PARTIAL FRACTION

this leads to a system of two equations:

$$1 = A + B.$$
$$0 = -4A + 2B.$$

This system can simply be solved using any method, or simpler using TI-83 as follows:
Press 2nd Matrix → Edit → Enter the size of Matrix (2x3), enter the coefficients of the first row $[1\ 1\ 1]$, then coefficients of the second row $[0\text{-}4\ 2]$ →Press second Quit.
Again Press 2nd Matrix → Math, then scroll down to rref(second Matrix) this gives rref(A), then press enter, and this gives the reduced matrix with answers: $A = 3/2$, and $B = -1/2$. Then the partial fraction decomposition is:

$$\frac{s}{s^2 - 2s - 8} = \frac{3/2}{s+2} - \frac{1/2}{s-4}.$$

2. repeated linear factors:
The rational appears as:

$$R(s) = \frac{p(s)}{(s-r)^n}.$$

Then the decomposed partial fraction is:

$$\frac{A_1}{(s-r)} + \frac{A_2}{(s-r)^2} + \ldots + \frac{A_n}{(s-r)^n}.$$

Where the numbers $A_1, A_2, ..., A_n$ are to de determined.

Example: Write the partial fraction decomposition of :

$$\frac{s+2}{s^3 - 2s^2 + s}$$

First we factor the denominator or $q(s)$:

$$q(s) = s(s^2 - 2s + 1) = s(s-1)^2$$

Then decompose the rational expression:

$$\frac{s+2}{s^3 - 2s^2 + s} = \frac{A}{s} + \frac{B}{(s-1)} + \frac{C}{(s-1)^2}.$$

The common denominator gives:

$$s + 2 = A(x-1)^2 + Bs(s-1) + Cs.$$

Solving for the numbers, A, B, C as before, gives:
$A = 2, B = -2, C = 3$.
Then the partial fraction decomposition is:

$$\frac{s+2}{s^3 - 2s^2 + s} = \frac{2}{s} - \frac{2}{(s-1)} + \frac{3}{(s-1)^2}.$$

3. quadratic factors: The rational appears as:

$$R(s) = \frac{p(s)}{[(s-a)^2 + b^2]^m}.$$

Then the partial fraction decomposition is:

$$\frac{A_1(s-a) + B_1 b}{[(s-a)^2 + b^2]} + \frac{A_2(s-a) + B_2 b}{[(s-a)^2 + b^2]^2} + \cdots$$
$$+ \frac{A_n(s-a) + B_n b}{[(s-a)^2 + b^2]^n}.$$

4.6. DECOMPOSITION OF PARTIAL FRACTION

Example: Write the partial fraction decomposition of

$$\frac{s^2 + 2s}{(s-1)(s^2 - 2s + 5)}$$

Notice the quadratic factor is irreducible, can be factored using completing the square method:

$$s^2 - 2s + 5 = \underbrace{s^2 - 2s + 1}_{(s-1)^2} + \underbrace{4}_{2^2} = (s-1)^2 + 2^2.$$

Then the partial fraction decomposition is:

$$\frac{s^2 + 2s}{(s-1)(s^2 - 2s + 5)} = \frac{A}{(s-1)} + \frac{B(s-1) + 2C}{(s-1)^2 + 2^2}.$$

As before, the common denominator gives:

$$s^2 + 2s = A(s^2 - 2s + 5) + [B(s-1) + 2C](s-1).$$

Solving for the numbers $A, B,$ and C in the following way:

$$O(s^2) : 1 = A + B.$$
$$O(s^1) : 2 = -2A - 2B + 2C.$$
$$O(s^0) : 0 = 5A + B - 2C.$$

Solving the system of 3-equations algebraically, or simply using TI-83:
{ Press 2nd Matrix → Edit → Enter the size of Matrix (3x4), enter the coefficients of the 3-rows: first row: [1 1 0 1], second row: [-2 -2 2 2],third row: [5 1 -2 0]→Press second Quit. Press 2nd Matrix → Math→ rref(second Matrix) → Enter } this gives the reduced matrix with answers:

$A = 3/4, B = 1/4$, and $C = 2$. Then the partial fraction decomposition is:

$$\frac{s^2 + 2s}{(s-1)(s^2 - 2s + 5)} = \frac{3/4}{(s-1)} + \frac{1/4(s-1) + 4}{(s-1)^2 + 2^2}.$$

Example: Find the inverse Laplace transform for the following:

$$\mathcal{L}^{-1}\{\frac{s-1}{s^2(s^2+9)}\}.$$

We will simplify the partial fraction first,

$$\frac{s-1}{s^2(s^2+9)} = \frac{A}{s} + \frac{B}{s^2} + \frac{Ds + E}{s^2 + 9}.$$

Solving for the constants A, B, D, and E gives:
$A = E = \frac{1}{9}$, and $B = D = -\frac{1}{9}$.

Substituting these constants we get,

$$\mathcal{L}^{-1}\{\frac{s-1}{s^2(s^2+9)}\} = \mathcal{L}^{-1}\{\frac{A}{s} + \frac{B}{s^2} + \frac{Ds + E}{s^2 + 9}\}.$$

$$= \frac{1}{9} - \frac{t}{9} - \frac{1}{9}cos3t + \frac{1}{27}sin3t.$$

Example - 4: IVP Solve the initial value problem:

$w'' - 2w' + 5w = e^t.$ *with* $w(0) = 0,\ w'(0) = 3.$

Applying the Laplace transform first:
$\mathcal{L}\{w'' - 2w' + 5w = e^t\}$. Then,

4.6. DECOMPOSITION OF PARTIAL FRACTION:

$$s^2W(s) - sw(0) - w'(0)$$
$$-2(sW(s) - w(0)) + 5W(s) = \frac{1}{(s+1)}.$$
$$s^2W - 3 - 2sW + 5W = \frac{1}{(s+1)}.$$
$$W(s^2 - 2s + 5) = \frac{1}{(s-1)} + 3.$$
$$W(s^2 - 2s + 5) = \frac{3s-2}{(s-1)}.$$
$$W(s) = \frac{3s-2}{(s-1)(s^2-2s+5)}.$$

Then the partial fraction decomposition is, as done above:

$$\frac{3s-2}{(s-1)(s^2-2s+5)} = \frac{A}{(s-1)} + \frac{B(s-1)+2C}{(s-1)^2+2^2}.$$

And the common denominator gives:

$$3s - 2 = A(s^2 - 2s + 5) + [B(s-1) + 2C](s-1).$$

Solving for the numbers $A, B,$ and C in the following way:

$$O(s^2) : 0 = A + B.$$
$$O(s^1) : 3 = -2A - 2B - 2C.$$
$$O(s^0) : -2 = 5A + B - 2C.$$

Solving the system of 3-equations algebraically, or simply using TI-83 gives:
$A = -5/4, B = 5/4,$ and $C = -3/2.$ Then the partial fraction decomposition is:

$$\frac{3s-2}{(s-1)(s^2-2s+5)} = \frac{-5/4}{(s-1)} + \frac{5/4(s-1)-3}{(s-1)^2+2^2}.$$

Now, step-2 is to apply the Laplace inverse transform:

$$\mathcal{L}^{-1}\{\frac{3s-2}{(s-1)(s^2-2s+5)}\} = \mathcal{L}^{-1}\{\frac{-5/4}{(s-1)} + \frac{5/4(s-1)-3}{(s-1)^2+2^2}\}.$$

$$\mathcal{L}^{-1}W(s) = -\frac{5}{4}\mathcal{L}^{-1}\{\frac{1}{(s-1)}\} + \frac{5}{4}\mathcal{L}^{-1}\{\frac{(s-1)-3}{(s-1)^2+2^2}\}.$$

$$w(t) = -\frac{5}{4}\mathcal{L}^{-1}\{\frac{1}{(s-1)}\} + \frac{5}{4}\mathcal{L}^{-1}\{\frac{(s-1)}{(s-1)^2+2^2}\}$$

$$-\frac{15}{8}\mathcal{L}^{-1}\{\frac{2}{(s-1)^2+2^2}\}.$$

$$\text{Then, } w(t) = -\frac{5}{4}e^t + \frac{5}{4}e^t\cos 2t - \frac{15}{8}e^t\sin 2t.$$

Example: IVP Solve the initial value problem:

$y'' + 4y' - 5y = te^t$. with $y(0) = 1$, $y'(0) = 0$.

Applying the Laplace transform first:

$$\mathcal{L}\{y'' + 4y' - 5y\} = te^t\}.$$
$$s^2Y(s) - sy(0) - y'(0)$$
$$+4(sY(s) - y(0)) - 5Y(s) = \frac{1}{(s+1)^2}.$$
$$s^2Y - s + 4sY - 4 - 5Y = \frac{1}{(s+1)^2}.$$
$$Y(s^2 + 4s - 5) = (s+4)\frac{1}{(s-1)^2}.$$
$$Y(s) = \frac{s^3 + 2s^2 - 7s + 5}{(s-1)^3(s+5)}.$$

Applying the partial fraction method gives:

$$\frac{s^3 + 2s^2 - 7s + 5}{(s-1)^3(s+5)} = \frac{A}{(s-1)^3} + \frac{B}{(s-1)^2} + \frac{C}{(s-1)} + \frac{D}{(s+5)}.$$

4.6. DECOMPOSITION OF PARTIAL FRACTION:

Solving for the constants: $A, B, C,$ and D will lead into a system of 4-equation which can easily be solved to give: $A = \frac{1}{6}$, $B = \frac{-1}{36}$, $C = \frac{181}{216}$, and $D = \frac{35}{216}$. Then applying the inverse Laplace transform to $Y(s)$ gives:

$$\mathcal{L}^{-1}Y(s) = \frac{1}{6}\mathcal{L}^{-1}\{\frac{1}{(s-1)^3}\} + \frac{-1}{36}\mathcal{L}^{-1}\{\frac{1}{(s-1)^2}\}$$
$$+\frac{181}{216}\mathcal{L}^{-1}\{\frac{1}{(s-1)}\} + \frac{35}{216}\mathcal{L}^{-1}\{\frac{1}{(s+5)}\}.$$

$$Or, \quad y(t) = \frac{1}{6}e^{3t} - \frac{1}{36}e^{2t} + \frac{181}{216}e^t + \frac{35}{216}e^{-5t}.$$

Example: IVP Solve the initial value problem:

$$u'' + u = cost. \quad with \quad u(0) = 0, \quad u'(0) = 0.$$

Applying the Laplace transform gives:

$$\mathcal{L}\{u'' + u = cost\}.$$
$$s^2 U(s) - su(0) - u'(0) + U = \frac{s}{(s^2+1)}.$$
$$U(s^2+1) = \frac{s}{(s^2+1)}.$$
$$U(s) = \frac{s}{(s^2+1)^2}.$$

Then applying the inverse Laplace transform to $U(s)$ gives:

$$\mathcal{L}^{-1}U(s) = \mathcal{L}^{-1}\{\frac{s}{(s^2+1)^2}\}.$$

4.7 Laplace for Unit-Step-Functions:

Earlier, in section (4.1), we have mentioned that Laplace Transform is used in Electrical engineering to solve circuit problems, but frequently they encounter functions which can be either "on" or "off" based on the forcing function acting on the circuit that could turn it off after a period of time. Then the function will be defined by different formula in different intervals. This is handled by the unit-step-function which is defined as:

$$u(t) = \begin{cases} 0 & \text{if } t < a \\ 1 & \text{if } a \leq t \end{cases}$$

Fig(6)   Unit-Step-Function at t=a

The unit-step-function for $f(t)$ shown in the graph is:

$$f(t) = 0u(t-0) + 1u(t-a) = u(t-a).$$

4.7. LAPLACE FOR UNIT-STEP-FUNCTIONS:

The following table shows some of the Laplace properties that are applied to the step-unit-functions:

$f(t)$	$F(s)$	
u(t-a)	$\dfrac{e^{-as}}{s}$	
$\mathcal{L}\{f(t-a)u(t-a)\}$	$e^{-as}F(s)$	Table(5)
$\mathcal{L}\{g(t)u(t-a)\}$	$e^{-as}\mathcal{L}\{g(t+a)\}$	
$\mathcal{L}^{-1}\{e^{-as}F(s)\}$	$f(t-a)u(t-a)\}$.	

Example-1: Graph the given step function:

$$f(t) = \begin{cases} 0 & \text{if } 0 \leq t < 1 \\ 2 & \text{if } t \geq 1 \end{cases}$$

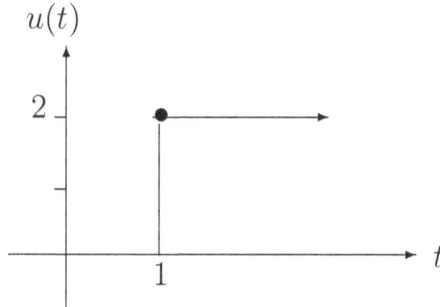

Fig(7) Unit-Step-Function at t=1

The unit-step-function for the above graph is:

$$f(t) = 0u(t-0) + 2u(t-1) = 2u(t-1).$$

Example-2: For the given graph write the step function, and find the Laplace Transform of the function:

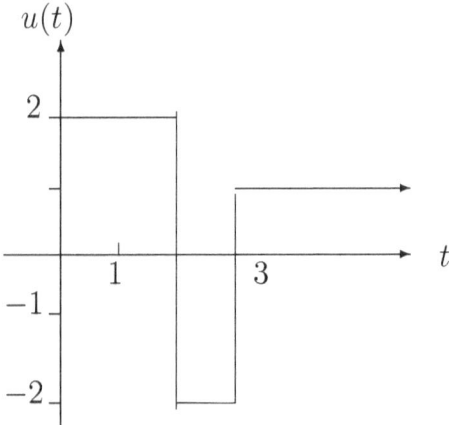

Fig(8) Unit-Step-Function

Solution: The step-function for the graph is:

$$f(t) = \begin{cases} 2 & \text{if } 0 \le t < 2 \\ -2 & \text{if } 2 \le t < 3 \\ 1 & \text{if } t \ge 3 \end{cases}$$

And the function is:

$$f(t) = 2 - 4u(t-2) + 3u(t-3).$$

Applying Laplace transform gives:

$$\mathcal{L}\{f(t)\} = \mathcal{L}\{2\} - \mathcal{L}\{4u(t-2)\} + \mathcal{L}\{3u(t-3)\}.$$

$$F(s) = \frac{2}{s} - \frac{4e^{-2s}}{s} + \frac{3e^{-3s}}{s}.$$

$$F(s) = \frac{2 - 4e^{-2s} - 3e^{-3s}}{s}.$$

4.7. LAPLACE FOR UNIT-STEP-FUNCTIONS:

Example-3: Find the Laplace transform for the step function:
$$f(t) = \begin{cases} 4 & \text{if } t < 1 \\ -2 & \text{if } 1 \leq t < 5 \\ 5 & \text{if } t \geq 5 \end{cases}$$

Solution: First we will write the function in equation form as:
$$f(t) = 4 - 6u(t-1) + 7u(t-5).$$

Then apply the Laplace transform,
$$\mathcal{L}\{f(t)\} = \mathcal{L}\{4 - 6u(t-1) + 7u(t-5).$$
$$F(t) = 4\mathcal{L}\{1\} - 6\mathcal{L}\{u(t-1)\} + 7\mathcal{L}\{u(t-5)\}.$$

Then applying the following formula:
$$\mathcal{L}\{u(t-a)\} = \frac{e^{-as}}{s}.$$

$$We\ get: F(s) = \frac{4}{s} - 6\frac{e^{-s}}{s} + 7\frac{e^{-5s}}{s}.$$

$$Or,\ F(s) = \frac{4 - 6e^{-s} + 7e^{-5s}}{s}.$$

Example-4: For the given graph:
a. Write the unit step function $u(t)$.
b. Write the unit step function $f(t)$.
c. Find the Laplace Transform $\mathcal{L}\{f(t)\}$.

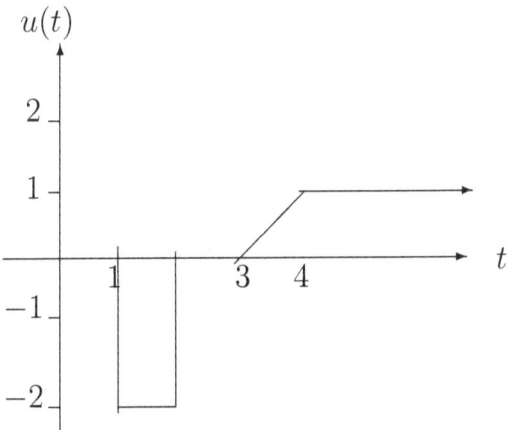

Fig(9) Unit-Step-Function

a. The unit step function intervals for $u(t)$ are:

$$u(t) = \begin{cases} 0 & \text{if } t < 1 \\ -2 & \text{if } 1 < t < 2 \\ 0 & \text{if } 2 < t < 3 \\ t & \text{if } 3 < t < 4 \\ 1 & \text{if } 4 < t \end{cases}$$

b. The unit step function $f(t)$ is:

$$\begin{aligned} f(t) &= 0u(t-0) - 2u(t-1) + 2u(t-2) + tu(t-3) \\ &+ (t-3)u(t-4). \\ f(t) &= -2u(t-1) + 2u(t-2) + tu(t-3) \\ &+ (t-3)u(t-4). \end{aligned}$$

c. Applying the Laplace Transform (using the properties given in table(5))gives:

4.7. LAPLACE FOR UNIT-STEP-FUNCTIONS:

$$\mathcal{L}\{f(t)\} = -\underbrace{\mathcal{L}\{2u(t-1)\}}_{\frac{2e^{-s}}{s}} + \underbrace{\mathcal{L}\{2u(t-2)\}}_{\frac{2e^{-2s}}{s}}$$

$$+ \underbrace{\mathcal{L}\{tu(t-3)\}}_{e^{-3s}\mathcal{L}(t+3)} + \underbrace{\mathcal{L}\{(t-3)u(t-4)\}}_{e^{-4s}\mathcal{L}(t+4-3)}.$$

$$F(s) = -\frac{2e^{-s}}{s} + \frac{2e^{-2s}}{s} + e^{-3s}\mathcal{L}\{(t+3)\}$$

$$+ e^{-4s}\mathcal{L}\{(t+1)\}.$$

$$= -\frac{2e^{-s}}{s} + \frac{2e^{-2s}}{s} + e^{-3s}(\frac{1}{s^2} + \frac{3}{s})$$

$$+ e^{-4s}(\frac{1}{s^2} + \frac{1}{s}).$$

Then, $F(s) = \frac{1}{s}(2e^{-2s} - 2e^{-s} + e^{-3s} + e^{-4s})$

$$+ \frac{1}{s^2}(e^{-3s} + e^{-4s}).$$

Example-5:
The non homogeneous differential equation with $g(t)$ given as a unit step function, graph and find the Laplace Transform $F(s)$ for the function:

$$y'' + 4y = g(t).$$

Where, $g(t)$ is given as:

$$g(t) = \begin{cases} 1 & \text{if } t < 1 \\ -1 & \text{if } 1 < t < 2 \\ 0 & \text{if } t \geq 2 \end{cases}$$

And initial values are given as: $y(0) = 0, y'(0) = 0$.

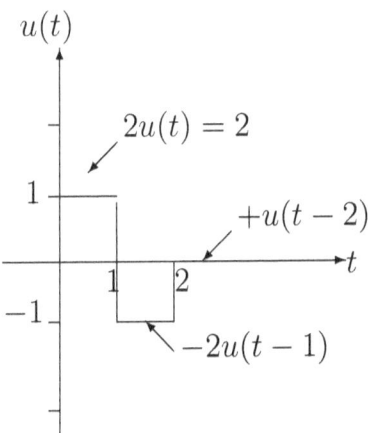

Fig(10)   Unit-Step-Function

The complete function for the problem is:

$$f(t) \;=\; 1 - 2u(t-1) + u(t-2).$$

Then the non homogeneous DE in terms of $g(t)$ is:

$$y'' + 4y \;=\; 1 - 2u(t-1) + u(t-2).$$

Applying Laplace Transform gives:

$$\mathcal{L}\{y''\} + \mathcal{L}\{4y\} \;=\; \mathcal{L}\{1\} - \mathcal{L}\{2u(t-1)\}$$
$$+ \; \mathcal{L}\{u(t-2)\}.$$
$$s^2 Y - sy(0) - y'(0) \;=\; \frac{1}{s} - \frac{2e^{-s}}{s} + \frac{e^{-2s}}{s}.$$

Applying the initial values and simplifying gives:

$$Y(s) = \frac{1}{s(s^2+4)} - \frac{2e^{-s}}{s(s^2+4)} + \frac{e^{-2s}}{s(s^2+4)}.$$

4.7. LAPLACE FOR UNIT-STEP-FUNCTIONS:

Note: The Laplace of Trigonometric functions require some trigonometric rules as shown in the following example:

$$\mathcal{L}\{cost\, u(t-\pi)\} \;=\; e^{-\pi s}\mathcal{L}\{cos(t+\pi)\}.$$

$$= e^{-\pi s}\mathcal{L}\{cost\, \underbrace{cos\pi}_{-1} - sint\, \underbrace{sin\pi}_{0}\}.$$

$$= e^{-\pi s}\mathcal{L}\{-cost\}.$$

$$= e^{-\pi s}(\frac{-s}{s^2+1}).$$

And similarly for the sine function:

$$\mathcal{L}\{sint\, u(t-\pi)\} \;=\; e^{-\pi s}\mathcal{L}\{sin(t+\pi)\}.$$

$$= e^{-\pi s}\mathcal{L}\{sint\, \underbrace{cos\pi}_{-1} + cost\, \underbrace{sin\pi}_{0}\}.$$

$$= e^{-\pi s}\mathcal{L}\{-sint\}.$$

Then, $\mathcal{L}\{sint\, u(t-\pi)\} \;=\; e^{-\pi s}(\dfrac{-1}{s^2+1}).$

Applications of Laplace Transform to Unit-Step-Functions

Electrical Circuits

As we have mentioned that Laplace Transform is used in Electrical engineering to solve circuit problems,and here we will present some examples: In the electrical circuits the following devices are used with symbols:
E - measured in Volts = the electro motive force (Battery).
L - measured in Henry = Inductance.
R - measured in ohms = resistance.
I- measured in Amperes = Current.
C- measured in Farads = Capacitance.

According to Faraday's Law: 1. The voltage drop across $L = E_L$, and it is promotional to the rate of change of the current I.

$$E_L = L\frac{dI}{dt}. \tag{4.5}$$

2. The voltage drop across the capacitor C is inversely proportional to the electric charge q:

$$E_C = \frac{q}{C}. \tag{4.6}$$

3. The Voltage drop across the resistor is proportional to the current I passing through R.

$$E_R = RI. \tag{4.7}$$

Then using **Kirchhoff's Conservation of Energy**, we get the total voltage applied to the following circuits at any time t:

4.7. LAPLACE FOR UNIT-STEP-FUNCTIONS:

A. If the circuit is RLC-Circuit, then the total voltage applied is:

$$E_R + E_L + E_C = E(t). \tag{4.8}$$

$$RI + L\frac{dI}{dt} + \frac{q}{C} = E(t). \tag{4.9}$$

But $I = \frac{dq}{dt}$, then equation (4.7) become:

$$L\frac{d^2q}{dt^2} + R\frac{dq}{dt} + \frac{1}{C}q = E(t). \tag{4.10}$$

Differentiating equation (4.10) twice with respect to t gives:

$$L\frac{d^3q}{dt^3} + R\frac{d^2q}{dt^2} + \frac{1}{C}\frac{dq}{dt} = \frac{dE}{dt}. \tag{4.11}$$

$$\tag{4.12}$$

$$L\frac{d^2}{dt^2}(\frac{dq}{dt}) + R\frac{d}{dt}(\frac{dq}{dt}) + \frac{1}{C}\frac{dq}{dt} = \frac{dE}{dt}. \tag{4.13}$$

$$\tag{4.14}$$

$$L\frac{d^2I}{dt^2} + R\frac{dI}{dt} + \frac{1}{C}I = \frac{dE}{dt}. \tag{4.15}$$

Equation (4.15) is the differential equation of the RLC-Circuit.

B. If the circuit is RL-Circuit, then the total voltage applied is:

$$E_L + E_R = E(t) \tag{4.16}$$

$$L\frac{dI}{dt} + RI = E(t) \tag{4.17}$$

$$L\frac{d^2q}{dt^2} + R\frac{dq}{dt} = E(t) \tag{4.18}$$

$$L\frac{dI}{dt} + RI = E(t) \tag{4.19}$$

If $R = 1$, and $L = 1$. Then, $\frac{dI}{dt} + I = E(t) \tag{4.20}$

Equation (4.20) is the differential equation of RL-Circuit.

Example-6: RL-Circuit(IVP)

Consider the initial value problem for the RL-Electric Circuit, with $R = L = 1$, and $I(0) = 0$, and the source voltage given as:

$$g(t) = \begin{cases} 2 & \text{if } 0 \leq t < 1 \\ -1 & \text{if } 1 \leq t < 2 \\ 0 & \text{if } t \geq 2 \end{cases}$$

Find the current $I(t)$ in the circuit.

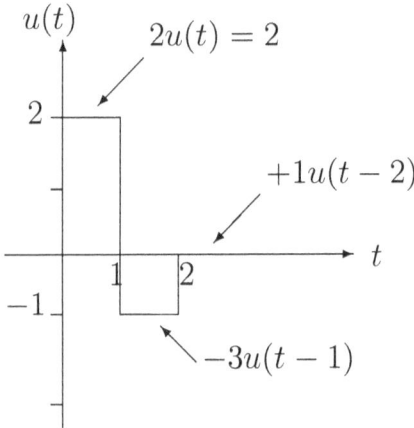

Fig(11) Unit-Step-Function

Solution: The differential equation for this circuit is :

$$\frac{dI}{dt} + I = E(t). \tag{4.21}$$

The source voltage is the unit step function given in the graph:

$$E(t) = 2 - 3u(t-1) + u(t-2).$$

4.7. LAPLACE FOR UNIT-STEP-FUNCTIONS:

Substituting in equation (4.21) gives:

$$\frac{dI}{dt} + I = 2 - 3u(t-1) + u(t-2).$$

Then applying the Laplace transform gives:

$$\mathcal{L}\{\frac{dI}{dt} + I\} = \mathcal{L}\{2 - 3u(t-1) + u(t-2)\}.$$

$$sI' - I(0) + I = \mathcal{L}\{2\} - 3\mathcal{L}\{u(t-1)\} + \mathcal{L}\{u(t-2)\}.$$

$$I(s+1) = \frac{2}{s} - \frac{3e^{-s}}{s} + e^{-2s}.$$

$$\text{Then, } I(s) = \frac{2}{s(s+1)} - \frac{3e^{-s}}{s(s+1)} + \frac{e^{-2s}}{(s+1)}.$$

$$I(s) = \frac{2}{s(s+1)} - \frac{3e^{-s}}{s(s+1)} + \frac{e^{-2s}}{(s+1)}$$

Applying the inverse Laplace transform to each term gives:

$$The\ first\ term : 2\mathcal{L}^{-1}\{\frac{1}{s^2+1}\} = 4sint.$$

$$The\ second\ term : 3\mathcal{L}^{-1}\{\frac{e^{-s}}{s^2+1}\} = -3sin(t-1)u(t-1)$$

$$The\ third\ term : \mathcal{L}^{-1}\{\frac{e^{-2s}}{s+1}\} = e^{-(t-1)}u(t-2).$$

Then, the solution is:

$$I(t) = 4sint - 3sin(t-1)u(t-1) + e^{-(t-1)}u(t-2).$$

Example-7: RLC-Circuit (IVP)

Consider an RLC-Circuit with $E(t) = sint$ volts. Find the current in the circuit at any time, where the constants are: $L = R = C = 1$ with initial values: $I(0) = -1 Amp, I'(0) = 8 Am/sec$.
Solution:
Using the differential equation for the RLC circuit or equation (4.15):

$$L\frac{d^2I}{dt^2} + R\frac{dI}{dt} + \frac{1}{C}I = \frac{dE}{dt}.$$

$$I'' + I' + I = \frac{d}{dt}(sint).$$

$$I'' + I' + I = cost.$$

Applying Laplace Transform:

$$\mathcal{L}\{I''\} + \mathcal{L}\{I'\} + \mathcal{L}\{I\} = \mathcal{L}\{cost\}.$$
$$s^2 I - sI(0) - I'(0) + sI - I(0) + I = \frac{s}{s^2 + 1^2}.$$
$$s^2 + s - 8 + sI + 1 + I = \frac{s}{s^2 + 1}.$$
$$I(s^2 + s + 1) = -(s - 7) + \frac{s}{s^2 + 1}.$$
$$I(s) = \frac{7 - s}{(s^2 + s + 1)} + \frac{s}{(s^2 + s + 1)(s^2 + 1)}.$$
$$I(s) = \frac{(7 - s)(s^2 + 25) + s}{(s^2 + s + 1)(s^2 + 1)}.$$
$$I(s) = \frac{s^3 - 7s^2 + 2s - 7}{(s^2 + 1)(s^2 + s + 1)}$$

4.7. LAPLACE FOR UNIT-STEP-FUNCTIONS:

Solving the partial fraction:

$$\frac{s^3 - 7s^2 + 2s - 7}{(s^2+1)(s^2+s+1)} = \frac{As+B}{s^2+1} + \frac{Cs+D}{(s+1/4)^2 + (\frac{\sqrt{3}}{2})^2}.$$

Gives,

$$-s^3 + 7s^2 + 7 = (As+B)(S^2+s+1)$$
$$+ (Cs+D)(^2+1).$$
$$O(s^3) : -1 = A+C.$$
$$O(S^2) : 7 = B+A+D.$$
$$O(s) : 0 = B+A+C.$$
$$O(s^0) : 7 = B+D.$$

Using Matrix with TI-83 to get the value of the constants:

$$\begin{bmatrix} 1 & 0 & 1 & 0 & -1 \\ 1 & 1 & 0 & 1 & 7 \\ 1 & 1 & 1 & 0 & 0 \\ 0 & 1 & 0 & 1 & 7 \end{bmatrix} \Rightarrow \begin{array}{l} A = 0 \\ B = 1 \\ C = -1 \\ D = 6 \end{array}$$

Substituting back and applying the inverse Laplace Transform gives:

$$I(s) = \mathcal{L}^{-1}\{\frac{1}{s^2+1}\} - \mathcal{L}^{-1}\{\frac{s-6}{(s+1/4)^2 + (\frac{\sqrt{3}}{2})^2}\}.$$

$$I(s) = \mathcal{L}^{-1}\{\frac{1}{s^2+1}\} - \mathcal{L}^{-1}\{\frac{s-6+1/4-1/4}{(s+1/4)^2 + (\frac{\sqrt{3}}{2})^2}\}.$$

$$Then, I(s) = sint - e^{-t/4}cos\frac{\sqrt{3}}{2}t - e^{-t/4}sin\frac{\sqrt{3}}{2}t.$$

4.8 Applications of Laplace Transform

In section (2.7) we have discussed the mixture problem using example of one-tank as application for first order differential equations, here we will discuss the mixture problem using example of two-tank-problem as an application for Laplace Transform.

Two-Tanks-Problem

The two tanks shown in the graph, each contains $40L$ of liquid, we need to find the amount of salt in each tank at time t, for the given information.

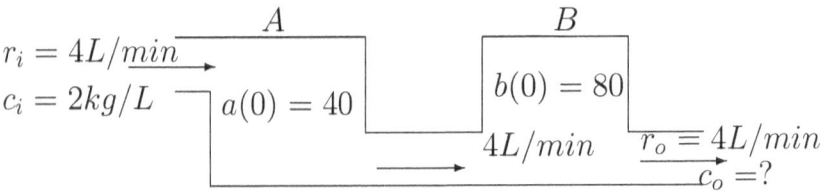

Fig(11) Mixing Problem with equal flow rates

- - - -

To solve two tank problem we apply the formula used before to each tank separately with the following considerations:

Tank-A :
Let, $A(t)$ = the amount of salt in tank A at time t.
$a(0) = 40$ kg of salt in tank A at time $t = 0$.
$r_i = 4L/min$.
$c_i = 2kg/L$.

4.8. APPLICATIONS OF LAPLACE TRANSFORM

$$\text{in} - \text{flow} \Rightarrow (r_i).(c_i) = 4L/min.2kg/L$$
$$= 8kg/min.$$

$r_o = 4L/min.$
$c_o = \frac{A(t)}{40}kg/L.$

$$\text{out} - \text{flow} \Rightarrow (r_o).(c_o) = 4L/min.\frac{A(t)}{40}kg/L.$$
$$= \frac{A(t)}{10}kg/min.$$

Applying the model formula of mixture problems:

$$\frac{dA}{dt} = (in\ flow) - (out\ flow).$$

$$\frac{dA}{dt} = 8 - \frac{A}{10}.$$

Tank-B Let,:
$B(t)$= the amount of the salt in tank B at time t.
$b(0) = 80$ kg of salt in tank B at time $t = 0$.
$r_i = 4L/min.$
$c_i = \frac{A(t)}{40}kg/L.$

$$\text{in} - \text{flow} \Rightarrow (r_i).(c_i) = 4L/min.\frac{A}{40}kg/L$$
$$- \frac{A(t)}{10}kg/L.$$

$r_o = 4L/min.$
$c_o = \frac{B(t)}{40kg/L}kg/L.$

$$\text{out} - \text{flow} \Rightarrow (r_o).(c_o) = 4L/min.\frac{B(t)}{40}kg/L$$
$$= \frac{B(t)}{10}.$$

CHAPTER 4.

As in Tank A, applying the rule of mixture problems:

$$\frac{dB}{dt} = (in\ flow) - (out\ flow).$$

$$\frac{dB}{dt} = \frac{A}{10} - \frac{B}{10}.$$

Now, the problem is : Solving the initial value problem of two-tanks in a system of two differential equations:

$$\frac{dA}{dt} = 8 - \frac{A}{10}.with,\ IV:\ a(0) = 40kg.$$

$$\frac{dB}{dt} = \frac{A}{10} - \frac{B}{10}, with\ IV:\ b(0) = 80.$$

Applying Laplace transform to the system of two equations, replacing both $\frac{dA}{dt}$, and $\frac{dB}{dt}$ with A', and B' respectively as follows:

$$A' = 8 - \frac{A}{10}.with,\ IV:\ a(0) = 40kg. \qquad (4.22)$$

$$B' = \frac{A}{10} - \frac{B}{10}, with\ IV:\ b(0) = 80. \qquad (4.23)$$

Applying Laplace transform, with IV, to the first equation (4.22):

$$sA - a(0) = 8\mathcal{L}\{1\} - \frac{A}{10}.$$

$$sA - 40 = \frac{8}{s} - \frac{A}{10}.$$

$$sA + \frac{A}{10} = 40 + \frac{8}{s}.$$

$$A(s + \frac{1}{10}) = 40 + \frac{8}{s}.$$

$$Then,\ A(s) = \frac{(40 + \frac{8}{s})}{(s + \frac{1}{10})}.$$

4.8. APPLICATIONS OF LAPLACE TRANSFORM

To find the amount of salt in tank-A at ant time A(t) we have to apply Laplace inverse to $A(s)$:

$$\mathcal{L}^{-1}\{A(s)\} = \mathcal{L}^{-1}\{\frac{(40+\frac{8}{s})}{(s+\frac{1}{10})}\}.$$

$$A(t) = \mathcal{L}^{-1}\{\frac{40}{(s+\frac{1}{10})}\} + \mathcal{L}^{-1}\{\frac{\frac{8}{s}}{(s+\frac{1}{10})}\}.$$

$$= \mathcal{L}^{-1}\{\frac{40}{(s+\frac{1}{10})}\} + \mathcal{L}^{-1}\{\underbrace{\frac{1}{s(s+\frac{1}{10})}}_{partial\ fraction}\}.$$

$$= \mathcal{L}^{-1}\{\frac{40}{(s+\frac{1}{10})}\} + \mathcal{L}^{-1}\{\frac{C}{s} + \frac{K}{(s+\frac{1}{10})}\}.$$

Where, C, and K are constants: $c = 10$, and $K = -10$, then,

$$A(t) = \mathcal{L}^{-1}\{\frac{40}{(s+\frac{1}{10})}\} + \mathcal{L}^{-1}\{\frac{10}{s}\} - \mathcal{L}^{-1}\{\frac{10}{(s+\frac{1}{10})}\}.$$

Then, the amount of salt in tank-A at any time-t is:

$$A(t) = 80 - 40e^{-t/10}. \qquad (4.24)$$

In a similar way, we apply Laplace transform, and IV, to equation(4.23):

$$sB - b(0) = \frac{A}{10} - \frac{B}{10}.$$

$$sB - 80 = \frac{A}{10} - \frac{B}{10}.$$

$$sB + \frac{B}{10} = 80 + \frac{A}{10}.$$

$$B(s + \frac{1}{10}) = 80 + \frac{A}{10}.$$

$$B(s) = \frac{(80 + \frac{A}{10})}{(s + \frac{1}{10})}.$$

To find the amount of salt in tank-B at any time B(t) we have to substitute A and Apply inverse Laplace transform to $B(s)$.

$$\mathcal{L}^{-1}\{B(s)\} = \mathcal{L}^{-1}\{\frac{(80+\frac{A}{10})}{(s+\frac{1}{10})}\}.$$

$$B(t) = \mathcal{L}^{-1}\{\frac{80}{(s+\frac{1}{10})}\} + \mathcal{L}^{-1}\{\frac{\frac{A}{10}}{s+\frac{1}{10}}\}.$$

Substituting $A(s)$ found as:

$$A(s) = \frac{(40+\frac{8}{s})}{(s+\frac{1}{10})}.$$

$$B(t) = \mathcal{L}^{-1}\{\frac{80}{(s+\frac{1}{10})}\} + \frac{1}{10}\mathcal{L}^{-1}\{\frac{40+\frac{8}{s}}{(s+\frac{1}{10})^2}\}.$$

$$= 80\mathcal{L}^{-1}\{\frac{1}{(s+\frac{1}{10})}\} + 4\mathcal{L}^{-1}\{\frac{1}{(s+\frac{1}{10})^2}\}$$

$$+ \frac{4}{5}\mathcal{L}^{-1}\{\underbrace{\frac{1}{s(s+\frac{1}{10})^2}}_{partial\ fraction}\}.$$

Simplifying the partial fraction:

$$\frac{1}{s(s+\frac{1}{10})^2} = \frac{K}{s} + \frac{L}{(s+\frac{1}{10})} + \frac{M}{(s+\frac{1}{10})^2}$$

Solving for the constants $K, L, and M$ gives: $K = 100$, $L = -100$, $M = -10$, substituting back and solving for the inverse Laplace gives:

$$\frac{1}{s(s+\frac{1}{10})^2} = \frac{100}{s} - \frac{100}{(s+\frac{1}{10})} - \frac{10}{(s+\frac{1}{10})^2}.$$

$$\text{Then, } B(t) = 80\mathcal{L}^{-1}\{\frac{1}{(s+\frac{1}{10})}\} + 4\mathcal{L}^{-1}\{\frac{1}{(s+\frac{1}{10})^2}\}$$
$$+ \frac{4}{5}\mathcal{L}^{-1}\{\frac{100}{s} - \frac{100}{(s+\frac{1}{10})} - \frac{10}{(s+\frac{1}{10})^2}\}.$$
$$B(t) = 80e^{-t/10} + 4te^{-t/10}$$
$$+ 40e^{-t/10} - 8te^{-t/10}.$$

Then the amount of salt in Tak-B at any time is:

$$B(t) = 80e^{-t/10} - 4te^{-t/10} \text{ kg}.$$

Solutions of the two tanks are:

$A(t) = 80 - 40e^{-t/10}$ kg of salt in tank-A

$B(t) = 80e^{-t/10} - 4te^{-t/10}$ kg of salt in tank-B.

4.9 Technical Writing

Students, in group of 3 or 4, will construct the following two-problems:
1. A one-tank problem with different flow rates.
2. An Electro-motive-Circuit.
Students will write the general formula for the system first, then solve the problem using Laplace Transform and its inverse.

4.10 Review Exercise

In problems 1-3 determine the Laplace-Transform for the given functions:

1. $e^{-3/4t}$.
2. $t^5 e^{-6t}$.
3. $8e^{3t}\cos 3t$.

In problems 4-6 determine the inverse Laplace Transform for the given functions:

4. $\dfrac{4}{(s+5)^2}$.

5. $\dfrac{3}{(s^2+4)}$.

6. $\dfrac{e^{-s}}{(s-1)(s-2)}$.

In problems 7-9 Solve the given initial value problem:

7. $y'' + 3y' + 4y = te^t$; $y(0) = 0$, $y'(0) = 1$.
8. $y'' + 4y = e^{3t}$; $y(0) = -1$, $y'(0) = 3$.
9. $y'' - 2y - 3 = 0$; $y(0) = 0$, $y'(0) = -5$.

10. Find the solution for the given IVP:

$$t^2 y'' + 2ty' - 2y = 0; \quad y(0) = 0, \, y'(0) = 0.$$

4.11 Review exercise Solutions

$$1. Using \; \mathcal{L}\{e^{at}\} = \dfrac{1}{s-a}.$$

4.11. REVIEW EXERCISE SOLUTIONS

Then, $\mathcal{L}\{e^{-3/4t}\} = \dfrac{1}{s+3/4}.$

2. Using $\mathcal{L}\{e^{at}t^n\} = \dfrac{n!}{(s-a)^{n+1}}.$

Then, $\mathcal{L}\{e^{-6t}t^5\} = \dfrac{5!}{(s+6)^6}.$

3. Using, $\mathcal{L}\{e^{at}\cos bt\} = \dfrac{s-a}{(s-a)^{@}+b^2}.$

Then, $\mathcal{L}\{8e^{3t}\cos 3t\} = \dfrac{8(s-3)}{(s-3)^2+3^2}.$

4. $\mathcal{L}^{-1}\{\dfrac{4}{(s+5)^2}\} = e^{-st}t.$

5. $\mathcal{L}^{-1}\{\dfrac{3}{(s^2+4)}\} = 3/2 \sin 2t.$

6. $\mathcal{L}^{-1}\{\dfrac{e^{-s}}{(s-1)(s-2)}\}.$

Here: Let $e^{-at} = e^{-s} \Rightarrow a = 1.$
And, $F(s) = \dfrac{1}{(s-1)(s-2)}.$
And, by partial fraction : $F(s) = \dfrac{-1}{s-1} + \dfrac{1}{s-2}.$

Then, $\mathcal{L}^{-1}\{\dfrac{e^{-s}}{(s-1)(s-2)}\}$
$= \mathcal{L}^{-1}\{\dfrac{-e^{-s}}{(s-1)}\} + \mathcal{L}^{-1}\{\dfrac{e^{-s}}{(s-2)}\}.$
$= [-e^{(t-1)} + e^{2(t-1)}]u(t-1).$

General Rule:

$$\mathcal{L}^{-1}\{\dfrac{e^{-ks}}{(s+\alpha)}\} = e^{-\alpha(t-k)} \cdot u(t-k).$$

7. Using Laplace transform:

$$\mathcal{L}\{y'' + 3y' + 4y\} = \mathcal{L}\{te^t\}.$$
$$\mathcal{L}\{y''\} = s^2Y - sy(0) - y'(0) = s^2Y - 1.$$
$$\mathcal{L}\{y'\} = sY - y(0) = sY.$$
$$\mathcal{L}\{y\} = Y.$$

Then substituting into the first equation gives:

$$s^2Y - 1 + 3sY + 4Y = \frac{1}{(s-1)^2}.$$

Solving for $Y(s)$ gives:

$$Y(s) = \frac{s^2 - 2s + 2}{(s-1)^2(s^2 + 3s + 4)}.$$

8. Applying Laplace transform:

$$\mathcal{L}\{y'' + 4y\} = \mathcal{L}\{e^{3t}\}.$$
$$s^2Y + 2 + 4Y = \frac{1}{(s-3)}.$$

Solving for $Y(s)$ gives:

$$Y(s) = \frac{(s^2 - 6s + 8)}{(s-3)(s^2 + 2^2)}.$$

Then the solution for the IVP is:

$$y(t) = \mathcal{L}^{-1}\{Y(s)\} = \mathcal{L}^{-1}\{\frac{(s^2 - 6s + 8)}{(s-3)(s^2 + 2^2)}\}.$$

9. Applying Laplace transform:

$$\mathcal{L}\{y'' - 2y - 3\} = 0.$$

$$s^2Y - 2Y + 2 = 0.$$

$$Y(s^2 - 2) = -2.$$

$$Y(s) = \frac{-2}{(s^2 - 2)}.$$

Then, $y(t) = \mathcal{L}^{-1}\{Y(s)\} = -\sqrt{2}\sin(\sqrt{2}t)$.

10. Applying Laplace Transform :

$$\mathcal{L}\{t^2 y'' + 2ty' - 2y = 0\}.$$

$$\mathcal{L}\{t^2 y''\} = (-1)^2 \frac{d^2 F}{ds^2}.$$

$$= s^2 Y'' + 4sY' + 2Y.$$

$$\mathcal{L}\{ty'\} = -\frac{dF}{ds} = -sY' - Y.$$

Then the equation becomes:

$$s^2 Y'' + 4sY' + 2Y + 2(-sY' - Y) - 2Y = 0.$$

$$s^2 Y'' + 2sY' - 2Y = 0$$

4.12 Chapter-4 Assignment

1. Find the Laplace Transform for the following expressions:

a. $\mathcal{L}\{e^{3x-5}\}$.

b. $\mathcal{L}\{e^{-2x}(x+3)\}$.

c. $\mathcal{L}\{t(\cos 4t - 2\sin 4t)\}$.

2. Find $f(t)$ for the following functions:

a. $\dfrac{1}{s^2} - \dfrac{2}{s^2-1}$.

b. $\dfrac{2s}{s^2-5s+6}$.

c. $\dfrac{3}{(2s+5)^2}$.

d. $\dfrac{1}{s^3}$.

4. Determine the partial fraction for the following rationals:

a. $\dfrac{s^2-20s+7}{(s-1)(s+3)(s+5)}$.

b. $\dfrac{7s^2+23s+30}{(s-2)(s^2+2s+5)}$.

5. Solve the Initial Value Problem ,using Laplace Transform:

a. $u'' + 6u' + 9u = 0$, with $u(0) = 1$, $u'(0) = 6$.

b. $u''' + 3u'' + 3u' + u = 0$. with $u(0) = -4$, $u'(0) = 4$, $u''(0) = -2$.

c. $u'' + 3tu' - 6u = 1$. with $u(0) = 0$. $u'(0) = 0$.

Chapter 5

Power Series For Solving
Ordinary Differential Equations

The differential equations that we have studied in the previous chapters possessed solutions that can be expressed in term of the functions that we are familiar with such as : Polynomials, Exponentials, Trigonometric, ...etc, but in real life there are many equations arise, that do not posses a solution which can be expressed in terms of a convenient function. In this chapter our goal is to obtain representations for solutions as a power series. Students studied polynomials in Intermediate Algebra, it was labeled $P(x)$, and expressed in the standard form as:

$$P(x) = a_0 + a_1 x + a_2 x^2 + ... + a_n x^n. \qquad (5.1)$$

Where, $a_0, a_1, ...a_n$ are constants, and n is any integer. The power series (5.1) is centered at $x = 0$.
Since we already know from Intermediate Algebra, that polynomials are: Continuous, Smooth, and differentiable

functions, then suppose we like to extend the expansion of $p(x)$ to $n \to \infty$, this will result in a series of infinite number of constants and in variable x with infinite power of $n's$, this series is called "**Power Series**", which is expressed as:

$$\sum_{n=0}^{\infty} a_n x^n = a_0 + a_1 x + a_2 x^2 + \ldots + a_n x^n. \quad (5.2)$$

The power series in (5.2) is called power series about $x = 0$, and it converges only at $x = 0$, or converges for all real numbers x.
If the power series is about some point "c", then equation (5.2) will be written as:

$$\sum_{n=0}^{\infty} a_n (x-c)^n = a_0 + a_1(x-c) + a_2(x-c)^2$$
$$+ \ldots + a_n(x-c)^n.$$

The above power series is centered at $x = c$, and always converges when $x = c$ which is considered as a general form. The power series may converge for some values of x and diverge for other values of x. An interval labeled "δ" and called radius of convergence is used as measuring devise for testing the convergence of a series as follows:
For series in (5.2):
1. $\sum_{n=0}^{\infty} a_n x^n$ converges absolutely if $\mid x \mid < \delta$.

2. $\sum_{n=0}^{\infty} a_n x^n$ diverges if $\mid x \mid > \delta$.

And for series that centers at $x = c$:

1. $\sum_{n=0}^{\infty} a_n (x-c)^n$ converges absolutely if $\mid x - c \mid < \delta$.

2. $\sum_{n=0}^{\infty} a_n (x-c)^n$ diverges if $\mid x - c \mid > \delta$.

The testing is done by the following ratio test:

$$\lim_{n\to\infty} \left| \frac{a_{n+1}}{a_n} \right| = \frac{1}{\delta}.$$

When, $\delta = 0$, then $\frac{1}{\delta} = \infty$. When $\delta = \infty$, then $\frac{1}{\delta} = 0$.

From this we deduce the followings:

1. if $\lim_{n\to\infty} \left| \frac{a_{n+1}}{a_n} \right| = \frac{1}{\delta} < 1 \Rightarrow$ Series converges.

2. if $\lim_{n\to\infty} \left| \frac{a_{n+1}}{a_n} \right| = \frac{1}{\delta} > 1 \Rightarrow$ Series diverges.

3. if $\lim_{n\to\infty} \left| \frac{a_{n+1}}{a_n} \right| = \frac{1}{\delta} = 1 \Rightarrow$ No conclusions are made.

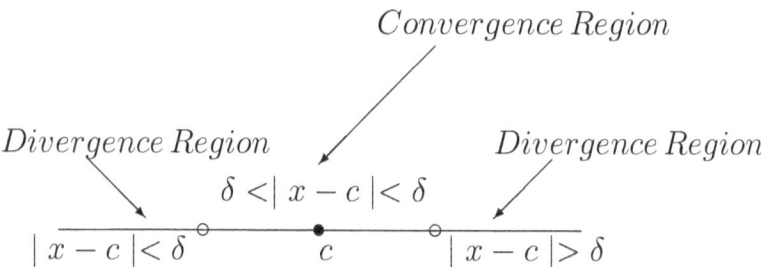

Fig(9) Convergence and divergence intervals

Example - 1: Apply the ratio test on the series:

$$\sum_{n=1}^{\infty} \frac{(x-5)^n}{3^n n}.$$

Solution:

$$\lim_{n\to\infty} \left\| \frac{\frac{(x-5)^{n+1}}{3^{n+1}(n+1)}}{\frac{(x-5)^n}{3^n n}} \right\| = \lim_{n\to\infty} \left\| \frac{(x-5)^{n+1}}{3^{n+1}(n+1)} \cdot \frac{3^n n}{(x-5)^n} \right\|.$$

$$= \lim_{n\to\infty} \left\| \frac{(x-5).n}{3(n+1)} \right\|.$$
$$= \frac{\|x-5\|}{3} \lim_{n\to\infty} \left\| \frac{n}{n+1} \right\|.$$
$$= \frac{\|x-5\|}{3}.$$

Then : $\lim_{n\to\infty} \sum_{n=1}^{\infty} \frac{(x-5)^n}{3^n n} = \frac{1}{3}\|x-5\|.$

1. If $\frac{1}{3}\|x-5\| < 1 \Rightarrow \|x-5\| < 3.$

    Or, $\Rightarrow -3 < x - 5 < 3.$

    $\Rightarrow 2 < x < 8.$

This means the series converges on the interval between 2, and 8 as shown in the fig.

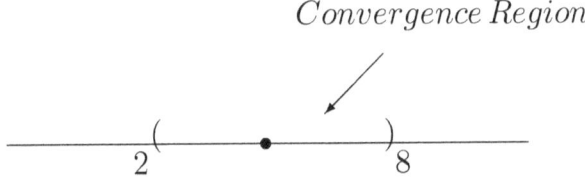
Convergence Region

2. If $\frac{1}{3}\|x-5\| > 1 \Rightarrow \|x-5\| > 3.$

    Or, $\Rightarrow x - 5 > 3 \text{ or, } x - 5 < -3.$

    $\Rightarrow x > 8, \text{ or } x < 2.$

This means the series diverges before 2, and after 8 as shown in the fig. below:

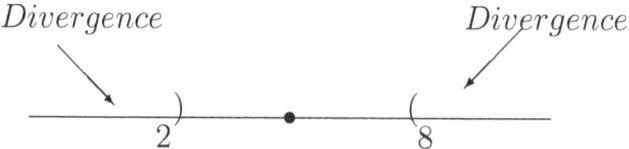

5.1 Differentiation, Integration of Power Series

Since we are going to use series to solve ordinary differential equations, then we need to know about the differentiation and integration of the power series. We will start with equation (5.1), and differentiate it once:

$$\{P(x)\}' = a_1 + 2a_2x + 3a_3x^2 + 4a_4x^3 + ... + na_nx^{n-1}$$
$$= \sum_{n=1}^{\infty} na_nx^{n-1}.$$

Then differentiating twice gives,

$$\{P(x)\}'' = 2a_2 + 6a_3x + 12a_4x^2 + ... + n(n-1)a_nx^{n-2}$$
$$= \sum_{n=2}^{\infty} n(n-1)a_nx^{n-2}.$$

Notice how the index of the Sum is changing based on the number of differentiation. In the same manner we will do the integration:

$$\int P(x)dx = \int a_0 dx + \int a_1x^2 dx + a_2x^2 + ... + \int a_nx^n dx$$

$$= a_0 x + a_1 \frac{x^2}{2} + a_2 \frac{x^3}{3} + \ldots + a_n \frac{x^{n+1}}{n+1}.$$

And integrating twice gives:

$$\int \{\int P(x)dx\}dx = \int a_0 x dx + \int \frac{a_1}{2} x^2 dx + \int \frac{a_2}{3} x^3 dx$$

$$+ \ldots + \int \frac{a_n x^{n+1}}{n+1} dx.$$

$$= a_0 \frac{x^2}{2} + a_1 \frac{x^3}{6} + a_2 \frac{x^4}{12}$$

$$+ \ldots + a_n \frac{x^{n+2}}{(n+1)(n+2)}.$$

5.2 Initial Value Problems of Power Series

It was Newton who considered the possibility of representing the solution of ordinary differential equations by infinite series. There are many types of infinite series, but the most familiar are the power series and Fourier series. The power series method is used to solve ordinary differential equations with constant coefficients, and non constant coefficients, and is applicable to both linear and nonlinear equations. In this section we will present some of these problems. We begin with first order linear differential equations written in the standard form as:

$$y'' + p(x)y' + q(x)y = 0. \tag{5.3}$$

Where $p(x)$, and $q(x)$ could be constants, or variable coefficient of polynomial, or trigonometric type. In the following examples, we will present some of these type of problems. When solving DE using series, the final solution is usually written in a general form , student has learned this in Calculus. Here we would rather visit some of those standard problems, that will help us with our solutions of Series.

1. $f(x) = 1 + x + \dfrac{x^2}{2!} + \dfrac{x^3}{3!} + \ldots = \sum\limits_{n=0}^{\infty} \dfrac{x^n}{n!}.$

2. $f(x) = -1 + \dfrac{x^2}{2!} - \dfrac{x^4}{4!} + \ldots = \sum\limits_{n=0}^{\infty} \dfrac{(-1)^{n+1} x^{2n}}{2n!}.$

3. $f(x) = -x + \dfrac{x^3}{3!} - \dfrac{x^5}{5!} + \ldots = \sum\limits_{n=0}^{\infty} \dfrac{(-1)^{n+1} x^{2n+1}}{(2n+1)!}.$

4. $e^x = 1 + x + \dfrac{x^4}{4!} + \dfrac{x^6}{6!} + \ldots = \sum\limits_{n=0}^{\infty} (-1)^{2n} \dfrac{x^n}{n!}.$

5. $e^{-x} = 1 - x + \dfrac{x^2}{2!} - \dfrac{x^3}{3!} + \ldots = \sum\limits_{n=0}^{\infty} (-1)^n \dfrac{x^n}{n!}.$

6. $\sin x = x - \dfrac{x^3}{3!} + \dfrac{x^5}{5!} - \dfrac{x^7}{7!} \ldots = \sum\limits_{n=0}^{\infty} (-1)^n \dfrac{x^{2n+1}}{(2n+1)!}.$

7. $\cos x = 1 - \dfrac{x^2}{2!} + \dfrac{x^4}{4!} - \dfrac{x^6}{6!} \ldots = \sum\limits_{n=0}^{\infty} (-1)^n \dfrac{x^{2n}}{(2n)!}.$

The above functions can easily be derived using Taylor Series expansion, here we will use Talor Series to derive the above last four functions.

we will use Taylor series centered at the origin which has the following form:

$$\sum\limits_{n=0}^{\infty} \dfrac{f^n(0)}{n!} x^n = f(0) + \dfrac{f'(0)}{1!} x + \dfrac{f''(0)}{2!} x^2 + \dfrac{f'''(0)}{3!} x^3 + \ldots$$

For the function : $f(x) = e^x$ the terms of Taylor series are:

$$f(0) = e^0 = 1.$$
$$f'(0) = 1.$$
$$f''(0) = 1.$$
$$f'''(0) = 1.$$

$$e^x = 1 + x + \frac{1}{2!}x^2 + \frac{1}{3!}x^3 + \ldots = \sum_{n=0}^{\infty}(-1)^{2n}\frac{x^n}{n!}.$$

For the function $f(x) = e^{-x}$:

$$f(0) = e^0 = 1.$$
$$f'(0) = -1.$$
$$f''(0) = 1.$$
$$f'''(0) = -1.$$

$$e^x = 1 - x + \frac{1}{2!}x^2 - \frac{1}{3!}x^3 + \ldots = \sum_{n=0}^{\infty}(-1)^n\frac{x^n}{n!}.$$

For the function $f(x) = sinx$:

$$f(0) = sin0 = 0.$$
$$f'(0) = cos0 = 1.$$
$$f''(0) = -sin0 = 0.$$
$$f'''(0) = -cos0 = -1.$$

$$sinx = x - \frac{x^3}{3!} + \frac{x^5}{5!} + \frac{x^5}{5!} + \ldots = \sum_{n=0}^{\infty}(-1)^n\frac{x^2n+1}{(2n+1)!}.$$

And the function $f(x) = \cos x$ is:

$$f(0) = \cos 0 = 1.$$
$$f'(0) = -\sin 0 = 0.$$
$$f''(0) = \cos 0 = 1.$$
$$f'''(0) = -\sin 0 = 0.$$

$$\cos x = 1 - \frac{x^2}{2!}x + \frac{x^4}{4!} - \frac{x^6}{6!} + \frac{x^8}{8!} + \ldots = \sum_{n=0}^{\infty} (-1)^n \frac{x^2 n}{(2n)!}.$$

Example - 2: Find the second derivative of the given function $f(x)$:

$$f(x) = 1 - \frac{x^2}{2!} + \frac{x^4}{4!} - \frac{x^6}{6!} \ldots = \sum_{n=0}^{\infty} (-1)^n \frac{x^{2n}}{2n!}.$$

$$f'(x) = -x + \frac{x^3}{3!} - \frac{x^5}{5!} + \ldots = \sum_{n=0}^{\infty} \frac{(-1)^{n+1} x^{2n+1}}{(2n+1)!}.$$

$$f''(x) = -1 + \frac{x^2}{2!} - \frac{x^4}{4!} + \ldots = \sum_{n=0}^{\infty} \frac{(-1)^{n+1} x^{2n}}{2n!}.$$

Example - 3: IVP Find the first four non-zero terms in a power series expansion about $x = 0$ for the general solution to the differential equation with constant coefficients:

$$y'' + 3y = 0. \qquad (5.4)$$

With initial values: $y(0) = 0$, and $y'(0) = 1$.

Solution: Equation (5.4) is a linear equation with constant coefficients, where : $p(x) = 3$, and $g(x) = 0$. We let the solution to be of the form $y =$ (5.2), and taking the first and second derivative gives:

$$y(x) = \sum_{n=0}^{\infty} a_n x^n. \qquad (5.5)$$

$$y'(x) = \sum_{n=1}^{\infty} n a_n x^{n-1}. \qquad (5.6)$$

$$y''(x) = \sum_{n=2}^{\infty} n(n-1) a_n x^{n-2}. \qquad (5.7)$$

Substituting (5.5 – 5.7) into (5.4) gives:

$$\sum_{n=2}^{\infty} n(n-1) a_n x^{n-2} + 3 \sum_{n=0}^{\infty} a_n x^n = 0. \qquad (5.8)$$

Now we will try to use the summation property to equate the exponents of the variable x, as follows:
Let $k = n-2 \to n = k+2$, on the first term, and replace n with k on the second term:

$$\underbrace{\sum_{k=0}^{\infty} (k+2)(k+1) a_{k+2} x^k}_{k=n-2} + 3 \underbrace{\sum_{k=0}^{\infty} a_k x^k}_{n \leftrightarrow k} = 0.$$

Then the coefficients of x^k gives:

$$(k+2)(k+1) a_{k+2} + 3 a_k = 0. \; for \; all \; k \geq 0.$$

And the "Recurrence" relation for the problem gives:

$$a_{k+2} = \frac{-3}{(k+1)(k+2)} a_k. \qquad (5.9)$$

Then writing the first four-nonzero-terms using relation (5.9),

$$k = 0 \to a_2 = \frac{-3a_0}{2}.$$
$$k = 1 \to a_3 = \frac{-3a_1}{6} = \frac{-a_1}{2}.$$
$$k = 2 \to a_4 = \frac{-3a_2}{12} = \frac{-3a_2}{12} = \frac{3a_0}{8}.$$
$$k = 3 \to a_5 = \frac{-3a_3}{20} = \frac{3a_1}{40}.$$
$$k = 4 \to a_6 = \frac{-3a_4}{30} = \frac{-3a_0}{80}.$$
$$k = 5 \to a_7 = \frac{-3a_5}{42} = \frac{-3a_1}{560}.$$
$$k = 6 \to a_8 = \frac{-3a_6}{56} = \frac{9a_0}{4480}.$$
$$k = 7 \to a_9 = \frac{-3a_7}{72} = \frac{a_1}{8960}.$$

Then the solution written in expansion form is:

$$y(x) = a_0 + a_1 x + a_2 x^2 + a_3 x^3 + a_4 x^4 ...$$
$$y'(x) = a_1 + 2a_2 x + 3a_3 x^2 + 4a_4 x^3 ...$$

Applying the given initial values: $y(0) = 0 = a_0$, and $y'(0) = 1 = a_1$. This means we have to drop out all the terms that deals with $a_0 = 0$. Then the solution is:

$$y(x) = x - \frac{x^3}{2} + \frac{3x^5}{40} - \frac{3x^7}{560} + \frac{3x^9}{13440} + ...$$

$$y(x) = x - \frac{3x^3}{6} + \frac{9x^5}{120} - \frac{27x^7}{5040} + \frac{81x^9}{362880} ...$$

$$y(x) = x - \frac{3x^3}{6} + \frac{9x^5}{120} - \frac{27x^7}{5040} + \frac{81x^9}{362880} ...$$

Then the general solution can be written as:

$$y(x) = x - \frac{3x^3}{3!} + \frac{9x^5}{5!} - \frac{27x^7}{7!} \cdots = \sum_{n=0}^{\infty}(-1)^n \frac{(3)^n x^{2n+1}}{(2n+1)!}.$$

Comparing with the formula in (6), the solution can be written as:

$$y(x) = sin\sqrt{3}x.$$

To check our solution in this example, we tempt to solve the problem using another method: Suppose the solution is of the form $y = e^{rx}$, then the general solution will be:

$$y(x) = c_1 e^{r_1 x} + c_2 e^{r_2 x}.$$

Differentiation y twice and Substituting :

$$y = e^{rx}.$$
$$y' = re^{rx}.$$
$$y'' = r^2 e^{rx}.$$

Substituting in (5.4) gives the auxiliary equation : $r^2 + 3 = 0$ with $r = \pm\sqrt{3}i$. Then the general solution is:

$$y(x) = e^{rx} = c_1 e^{r_1 x} + c_2 e^{r_2 x}.$$
$$= c_1 cos\sqrt{3}x + c_2 sin\sqrt{3}x.$$

Applying the given initial values leads to : $c_1 = 0$, and $c_2 = \frac{1}{\sqrt{3}}$. This proves that the general solution is $sin\sqrt{3}x$.

Example - 4: Find the power series expansion for the differential equation centered at $x = 0$.

$$y' - xy = 0. \qquad (5.10)$$

Solution: Using the solution of power series centered at the origin, and differentiate once gives:

$$y(x) = \sum_{n=0}^{\infty} a_n x^n.$$

$$y'(x) = \sum_{n=1}^{\infty} n a_n x^{n-1}.$$

Then, substitute back in (5.10):

$$y' - xy = 0.$$

$$\sum_{n=1}^{\infty} n a_n x^{n-1} - x \sum_{n=0}^{\infty} a_n x^n = 0.$$

$$\underbrace{\sum_{n=1}^{\infty} n a_n x^{n-1}}_{let\ k=n-1 \rightarrow n=k+1} - \underbrace{\sum_{n=0}^{\infty} a_n x^{n+1}}_{let\ k=n+1 \rightarrow n=k-1} = 0.$$

$$\sum_{k=0}^{\infty} (k+1) a_{k+1} x^k - \sum_{k=1}^{\infty} a_{k-1} x^k = 0.$$

$$a_1 + \sum_{k=1}^{\infty} (k+1) a_{k+1} x^k - \sum_{k=1}^{\infty} a_{k-1} x^k = 0.$$

Then taking the coefficients gives:

$$O(x^0) : a_1 = 0.$$
$$O(x^k) : (k+1) a_{k+1} - a_{k-1} = 0.$$

Then for $k \geq 1$ we get the following recurrence relation:

$$a_{k+1} = \frac{a_{k-1}}{(k+1)}.$$

Substituting the values for $k \geq 1$ gives:

$$k = 1 \rightarrow a_2 = \frac{a_0}{2}.$$

$$k = 2 \rightarrow a_3 = \frac{a_1}{3} = 0.$$

$$k = 3 \rightarrow a_4 = \frac{a_2}{4} = \frac{a_0}{8}.$$

$$k = 4 \rightarrow a_5 = \frac{a_3}{5} = 0.$$

$$k = 5 \rightarrow a_6 = \frac{a_4}{6} = \frac{a_0}{48}.$$

Then the power series is:

$$\begin{aligned} y(x) = \sum_{n=0}^{\infty} a_n x^n &= a_0 + a_1 x + a_2 x^2 + a_3 x^3 + a_4 x^4 + \dots \\ &= a_0 + \frac{1}{2} a_0 x^2 + \frac{1}{8} a_0 x^4 + \frac{1}{48} a_0 x^6 + \dots \\ \text{Then, } y(x) &= a_0 \sum_{n=0}^{\infty} \frac{(-1)^{2n}}{(n!) 2^n} x^{2n}. \end{aligned}$$

Example - 5:

Find the first four nonzero terms for the power series expansion of the given differential equation:

$$y'' - \sin x \, y = 0. \tag{5.11}$$

Solution: Equation (5.11) is a linear equation with variable coefficients, where : $p(x) = \sin x$. Let the solution to be of the form $y = 5.2$, and taking the first and second derivative gives:

$$y(x) = \sum_{n=0}^{\infty} a_n x^n. \tag{5.12}$$

$$y'(x) = \sum_{n=1}^{\infty} n a_n x^{n-1}. \tag{5.13}$$

$$y''(x) = \sum_{n=2}^{\infty} n(n-1) a_n x^{n-2}. \tag{5.14}$$

$$And, \sin x = \sum_{n=0}^{\infty}(-1)^n \frac{x^2n+1}{(2n+1)!}. \quad (5.15)$$

$$(5.16)$$

Substituting $(5.12 - 5.15)$ into (5.11) gives:

$$\sum_{n=2}^{\infty} n(n-1)a_n x^{n-2} - \{\sum_{n=0}^{\infty}(-1)^n \frac{x^2n+1}{(2n+1)!}\}\{\sum_{n=0}^{\infty} a_n x^n\} = 0.$$

Expanding the summation of each term gives:

$$(2a_2 + 6a_3 x + 12a_4 x^2 + 20a_5 x^3 + ...)$$
$$-(x - \frac{x^3}{6} + \frac{x^5}{120} + ...)(a_0 + a_1 x + a_2 x^2 + a_3 x^3 + ...) = 0.$$

Then collecting coefficients of x's gives:

$$O(x^0) : 2a_2 = 0 \to a_2 = 0.$$
$$O(x^1) : 6a_3 - a_0 = 0 \to a_3 = \frac{a_0}{6}.$$
$$O(x^2) : 12a_4 - a_1 = 0 \to a_4 = \frac{a_1}{12}.$$
$$O(x^3) : 20a_5 - \frac{a_0}{6} + a_2 = 0 \to a_5 = \frac{a_0}{120}.$$

Then the solution is:

$$y(x) = \sum_{n=0}^{\infty} a_n x^n = a_0 + a_1 x + a_2 x^2 + a_3 x^3 + a_4 x^4 + a_5 x^5 ...$$

$$= a_0 + a_1 x + \frac{a_0}{6} x^3 + \frac{a_1}{12} x^4 + \frac{a_0}{120} x^5 + ...$$

$$Or, \; y(x) = a_0(1 + \frac{x^3}{6} + \frac{x^5}{120} + ...) + a_1(x - \frac{x^4}{12} + ...)$$

Note: If, $y(x) = \sum_{n=0}^{\infty} a_n x^n.$

$$\begin{aligned}
Then, \; y^2 &= (\sum_{n=0}^{\infty} a_n x^n)^2. \\
&= (\sum_{n=0}^{\infty} a_n x^n)(\sum_{n=0}^{\infty} a_n x^n). \\
&= (a_0 + a_1 x + a_2 x^2 + a_3 x^3 + ...) \\
&\quad (a_0 + a_1 x + a_2 x^2 + a_3 x^3 + ...). \\
&= a_0 a_0 + (a_0 a_1 + a_1 a_0) x \\
&\quad + (a_0 a_2 + a_1 a_1 + a_2 a_0) x^2 + ...
\end{aligned}$$

Example - 6:
Find the first four nonzero terms for the power series expansion of the given differential equation:

$$y' - y^2 = 0. \tag{5.17}$$

Let the solution be of the form 5.2, and taking the first derivative gives:

$$y = \sum_{n=0}^{\infty} a_n x^n.$$

$$y^2 = (\sum_{n=0}^{\infty} a_n x^n)^2.$$

$$y'(x) = \sum_{n=1}^{\infty} n a_n x^{n-1}.$$

Then substituting in 5.17 gives:

$$y' - y^2 = 0.$$

$$\sum_{n=1}^{\infty} n a_n x^{n-1} - (\sum_{n=0}^{\infty} a_n x^n)^2 = 0.$$

$$(a_1 + 2a_2 x + 3a_3 x^2 + 4a_4 x^3 + \ldots)$$
$$-(a_0 + a_1 x + a_2 x^2 + a_3 x^3 + \ldots)$$
$$(a_0 + a_1 x + a_2 x^2 + a_3 x^3 + \ldots) = 0.$$

$$a_1 + 2a_2 x + 3a_3 x^2 + 4a_4 x^3 + \ldots$$
$$- a_0 a_0 - (a_0 a_1 + a_1 a_0) x$$
$$- (a_0 a_2 + a_1 a_1 + a_2 a_0) x^2 + \ldots = 0$$

Collecting coefficients gives:

$$O(x^0) : a_1 - a_0^2 = 0 \rightarrow a_1 = a_0^2.$$
$$O(x^1) : 2a_2 - a_0 a_1 - a_1 a_0 = 0 \rightarrow a_2 = a_0^3.$$
$$O(x^2) : 3a_3 - a_0 a_2 - a_1 a_1 - a_2 a_0 = 0 \rightarrow a_3 = a_0^4.$$

Then the solution is:

$$y(x) = a_0 + a_1 x + a_2 x^2 + a_3 x^3 + a_4 x^4 + \ldots$$
$$= a_0 + a_0^2 x + a_0^3 x^2 + a_0^4 x^4 + \ldots$$
$$y(x) = \sum_{n=0}^{\infty} (a_0)^{n+1} x^n.$$

5.3 Technical Writing

For the given equation :

$$(1-x^2)y'' - y' + y = \sin x.$$

Use a solution of the form :

$$y(x) = \sum_{n=0}^{\infty} a_n x^n.$$

And the expansion of $\sin x$ to solve the differential equation showing all the steps in detail.

5.4 Review Exercise

In problems 1 - 3 Find at least the first four nonzero terms in a power series expansion about $x = 0$ for a general solution for the given equations:

$$1. u'' + t^2 u' - 2u = 0.$$

$$2. u' - u = 0.$$

$$3. u' - 2tu = 0.$$

5.5 Review Exercise Solutions

1. Let the solution of power series type be:

$$u(t) = a_0 + a_1 t + a_2 t^2 + \ldots = \sum_{n=0}^{\infty} a_n t^n.$$

5.5. REVIEW EXERCISE SOLUTIONS

Differentiating, with respect to t, twice gives:

$$u'(t) = a_1 + 2a_2t + 3a_3t^2 + \ldots = \sum_{n=1}^{\infty} na_n t^{n-1}.$$

$$u''(t) = 2a_2 + 6a_3t + 12a_4t^2 + \ldots = \sum_{n=2}^{\infty} n(n-1)t^{n-2}.$$

Substituting back into the original DE:

$$\sum_{n=2}^{\infty} n(n-1)a_n t^{n-2} + t^2 \sum_{n=1}^{\infty} na_n t^{n-1} - 2\sum_{n=0}^{\infty} a_n t^n = 0.$$

Now, we have to shift the indicies:

$$First - Term: \sum_{n=2}^{\infty} n(n-1)a_n t^{n-2} \Rightarrow Let\ k = n-2.$$

$$Then, \Rightarrow n = k+2.$$

$$And, \sum_{n=2}^{\infty} n(n-1)a_n t^{n-2} \Rightarrow \sum_{k=0}^{\infty} (k+2)(k+1)a_{k+2} t^k.$$

$$Second-Term: t^2 \sum_{n=1}^{\infty} na_n t^{n-1} = \sum_{n=1}^{\infty} na_n t^{n+1}. Let\ k = n+1.$$

$$Then \Rightarrow n = k-1.$$

$$And, \sum_{n=1}^{\infty} na_n t^{n+1} \Rightarrow \sum_{k=0}^{\infty} (k-1)a_{k-1} t^k.$$

$$Third - term: \sum_{n=0}^{\infty} a_n t^n \Rightarrow \sum_{k=0}^{\infty} a_k t^k. Where, k = n.$$

Substituting these expansions back into the original differential equation:

$$\sum_{k=0}^{\infty} (k+2)(k+1)a_{k+2} t^k + \sum_{k=0}^{\infty} (k-1)a_{k-1} t^k - 2\sum_{k=0}^{\infty} a_k t^k = 0.$$

Collecting, all coefficients of t^k gives:

$$(k+2)(k+1)a_{k+2} + (k-1)a_{k-1} - 2a_k = 0.$$

From this we get : $a_{k+2} = \dfrac{2a_k - (k-1)a_{k-1}}{(k+1)(k+2)}$, $k \geq 1$.

The last recursion formula gives:

$$k = 1 \Rightarrow a_3 = 1/3 a_1.$$
$$k = 2 \Rightarrow a_4 = 1/6 a_2 - 1/12 a_1.$$
$$k = 3 \Rightarrow a_5 = 1/30 a_1 - 1/10 a_2.$$
$$k = 4 \Rightarrow a_6 = 1/90 a_2 - 7/180 a_1.$$

.

.

Then the solution is:

$$u(t) = \sum_{n=0}^{\infty} a_n t^n = a_0\{1 + t^2 + \ldots\} + a_1\{t + 1/3 t^3 + \ldots\}.$$

2. Let the solution of power series type be:

$$u(t) = a_0 + a_1 t + a_2 t^2 + \ldots = \sum_{n=0}^{\infty} a_n t^n.$$

Differentiating once and substituting back into the Original DE:

$$u'(t) = a_1 + 2a_2 t + 3a_3 t^2 + \ldots = \sum_{n=1}^{\infty} n a_n t^{n-1}.$$

The DE becomes : $\sum_{n=1}^{\infty} n a_n t^{n-1} - \sum_{n=0}^{\infty} a_n t^n = 0.$

$First - Term : \sum_{n=1}^{\infty} n a_n t^{n-1}.Let\ k = n - 1.$

5.5. REVIEW EXERCISE SOLUTIONS

$$Then \Rightarrow n = k + 1.$$

$$And, \ \sum_{n=1}^{\infty} na_n t^{n+1} \Rightarrow \sum_{k=0}^{\infty} (k+1)a_{k+1} t^k.$$

$$Second-term: \sum_{n=0}^{\infty} a_n t^n \Rightarrow \sum_{k=0}^{\infty} a_k t^k. \ Where, \ k = n.$$

Substituting, we get:

$$\sum_{k=0}^{\infty} (k+1)a_{k+1} t^k - \sum_{k=0}^{\infty} a_k t^k = 0.$$

$$From \ this \ we \ get \ ; a_{k+1} = \frac{a_k}{k+1}. \ \ k \geq 0.$$

The last expression gives:

$$k = 0 \Rightarrow a_1 = a_0.$$
$$k = 1 \Rightarrow a_2 = 1/2 a_0.$$
$$k = 2 \Rightarrow a_3 = 1/6 a_0.$$
$$k = 3 \Rightarrow a_4 = 1/24 a_0.$$
$$k = 4 \Rightarrow a_5 = 1/120 a_0.$$

.
.

Then the General solution for the given differential equation is:

$$u(t) = a_0 + a_1 t + a_2 t^2 + ... = \sum_{n=0}^{\infty} a_n t^n.$$

$$= a_0\{1 + \frac{t}{1!} + \frac{t^2}{2!} + \frac{t^3}{3!} + \frac{t^4}{4!}...\}.$$

Or, $u(t) = e^t$.

3. As in problems 1, and 2, let the solution be:

$$u(t) = a_0 + a_1 t + a_2 t^2 + ... = \sum_{n=0}^{\infty} a_n t^n.$$

Differentiating once and substituting back into the Original DE:

$$u'(t) = a_1 + 2a_2 t + 3a_3 t^2 + \ldots = \sum_{n=1}^{\infty} n a_n t^{n-1}.$$

The DE becomes : $\sum_{n=1}^{\infty} n a_n t^{n-1} - 2t \sum_{n=0}^{\infty} a_n t^n = 0.$

Or, $\sum_{n=1}^{\infty} n a_n t^{n-1} - 2 \sum_{n=0}^{\infty} a_n t^{n+1} = 0.$

First $-$ Term : $\sum_{n=1}^{\infty} n a_n t^{n-1}$. Let $k = n-1$.

$$\text{Then} \Rightarrow n = k+1.$$

And, $\sum_{n=1}^{\infty} n a_n t^{n+1} \Rightarrow \sum_{k=0}^{\infty} (k+1) a_{k+1} t^k.$

Second $-$ term : $\sum_{n=0}^{\infty} a_n t^{n+1} \Rightarrow$ Let $k = n+1.$

$$\text{Then} \Rightarrow n = k-1.$$

then, $\sum_{n=0}^{\infty} a_n t^{n+1} = \sum_{k=1}^{\infty} a_{k-1} t^k$

Substituting the DE becomes:

$$\sum_{k=0}^{\infty} (k+1) a_{k+1} t^k - 2 \sum_{k=1}^{\infty} a_{k-1} t^k = 0.$$

Shifting the first index we get:

$$a_1 + \sum_{k=1}^{\infty} (k+1) a_{k+1} t^k - 2 \sum_{k=1}^{\infty} a_{k-1} t^k = 0. \ k \geq 1.$$

This gives ; $a_1 = 0$, and from the coefficients of t^k we get the following relation:

$$a_{k+1} = \frac{2}{(k+1)} a_{k-1}.$$

Then the constants are:
$$k = 1 \Rightarrow a_2 = a_0.$$
$$k = 2 \Rightarrow a_3 = 2/3a_1.$$
$$k = 3 \Rightarrow a_4 = 1/2a_0.$$
$$k = 4 \Rightarrow a_5 = 4/15a_1.$$

.
.

then the solution is:
$$u(t) = a_0 + a_1 t + a_2 t^2 + \ldots = \sum_{n=0}^{\infty} a_n t^n.$$
And, $u(t) = a_0\{1 + t^2 + \ldots\} + a_1\{t + 2/3 t^3 + \ldots\}.$

5.6 Chapter-5 Assignment

1. Write at least the first 4-nonzero-terms of power series expansion about $x = 0$ for the general solution of the following differential equations:

   a) $y' + xy = 0.$
   b) $y'' - y = 0.$
   c) Use other methods to solve (b).

2. Write at least the first 4-nonzero-terms of power series expansion about $x = 0$ for the general solution of the following initial value problem:

$$x^2 y'' + xy' + xy = 0. \text{ with } y(0) = 0, \ y'(0) = 1.$$

3. Solve the initial value problem:

$$y'' + xy' - y = 0. \ y(0) = -1, \ y'(0) = 0.$$

4. Using Taylor Series, expand the given function and give the general form for the function: $f(x) = sin2x$.

5. Write the intervals of the unit-step function $f(t)$, graph the unit-step function, and use the Laplace and inverse transform to find the solution $f(t)$:

$$f(t) = -2 + 5u(t-3) - 3u(t-5) + 2u(t-6).$$

Chapter 6

Review from Calculus

Partial Derivatives and Chain Rule

In this section we will use partial derivative and describe its affect on functions. The process of partial derivative can be described by answering the following question:

Question: How does the volume (V) in a cylinder vary with respect to its radius (r), and height (h) in the equation:

$$V = \pi r^2 h \tag{6.1}$$

Answer: We hold one of the two variables fixed, and study the other variable, then reverse the process in terms of the other variable being fixed, this process is called **Partial Differentiation**.

> **Note:**
> Partial differentiation ∂ is different from regular differentiation d.

CHAPTER 6. REVIEW FROM CALCULUS

To differentiate a function of single variable $y = f(x)$ partially with respect to the independent variable x we write:
$$\frac{\partial y}{\partial x} = \frac{\partial f(x)}{\partial x}. \tag{6.2}$$

To differentiate a function of two - variables $z = f(x, y)$ partially with respect of the two independent variables x, and y, we have to do it in two steps with respect of the two variables separately:

1. Differentiation with respect to x:
$$\frac{\partial z}{\partial x} = \frac{\partial f(x, y\ fixed)}{\partial x}. \tag{6.3}$$

2. Differentiation with respect to y:
$$\frac{\partial z}{\partial y} = \frac{\partial f(x\ fixed, y)}{\partial y}. \tag{6.4}$$

Another symbol is used for partial differentiation as shown:
$$\frac{\partial z}{\partial x} = z_x.$$
$$\frac{\partial z}{\partial y} = z_y.$$

The formal definition of the differentiation is:
$$f_x(x, y) = lim_{\Delta x \to 0} \frac{f(x + \Delta x, y) - f(x, y)}{\Delta x}. \tag{6.5}$$

Application of Partial Differentiation:
Equations that involves partial differentiation are:
1. Heat Equations:
$$k\frac{\partial^2 u}{\partial x^2} = \frac{\partial u}{\partial t}, \quad k > 0 \tag{6.6}$$

Where, $k = constant$, $u = velocity$, $x = space$, and $t = time$. Equation (6.6) is first order in time, and second order in space, it is used in the theory of heat flow in a rod, or thin wire.

2. Wave Equation:

$$k\frac{\partial^2 u}{\partial x^2} = \frac{\partial^2 u}{\partial t^2}, \quad k > 0 \qquad (6.7)$$

This equation is twice differentiable in both time and space. It is used in physics.

3. Laplace Equation:

$$\frac{\partial^2 u}{\partial x^2} + \frac{\partial^2 u}{\partial y^2} + \frac{\partial^2 u}{\partial z^2} = 0. \qquad (6.8)$$

At steady-state (t=0) equation (6.7) becomes independent of time, and hence reduces to equation (6.8) the Laplacian.

Example- 1:
Find $f_x(x,y)$, and $f_y(x,y)$ for the given function:
$f(x,y) = x^2 - y^2$.

Solution:

$$\frac{\partial f(x,y)}{\partial x} = f_x(x, y\,fixed) = 2x.$$

$$\frac{\partial f(x,y)}{\partial y} = f_y(x\,fixed, y) = -2y.$$

Example- 2:
Find $f_x(x,y)$, and $f_y(x,y)$ for the given function:
$f(x,y) = \cos(2xy)$.

Solution:

$$\frac{\partial f(x,y)}{\partial x} = f_x(x, y\,fixed) = -2y\sin(2xy).$$

$$\frac{\partial f(x,y)}{\partial y} = f_y(x\,fixed, y) = -2x\sin(2xy).$$

Example- 3:
Find the value of $f_x(x,y)$, and $f_y(x,y)$ for the given function: $f(x,y) = xy^2 - 4x^2$ at the point $(3, -2)$.

Solution:

$$\frac{\partial f(x,y)}{\partial x} = f_x(x, y\,fixed) = y^2 - 8x = -20.$$

$$\frac{\partial f(x,y)}{\partial y} = f_y(x\,fixed, y) = 2xy = -12.$$

The derivative of vector-function $r(t)$ is given as:

$$r'(t) = \lim_{h \to 0} \frac{r(t+h) - r(t)}{h}. \qquad (6.9)$$

Provided that the limit exists. Also we can write:
If $r(t) = (x(t), y(t), z(t))$ and $x'(t), y'(t),$ and $z'(t)$ exist, then, $r'(t) = (x'(t), y'(t), z'(t))$.

Example- 4:
Find $r'(t)$ and find its values at $t = \pi/4$ for the vector function : $r(t) = (1/2\sin t, 3/2\cos t)$.

Solution:

$$\begin{aligned} r'(t) &= ((1/2\sin t)', (3/2\cos t)') \\ &= (1/2\cos t, -3/2\sin t) \end{aligned}$$

Then at $t = \frac{\pi}{4}$ gives $\Rightarrow$

$$r'(\pi/4) = (\frac{1}{2}\cos(\frac{\pi}{4}), \frac{-3}{2}\sin(\frac{\pi}{4}))$$
$$= (\frac{1}{2}\cdot\frac{1}{\sqrt{2}}, \frac{-3}{2}\cdot\frac{1}{\sqrt{2}})$$
$$= (\frac{1}{2\sqrt{2}}, \frac{1}{2\sqrt{2}}).$$

6.1 The Chain Rule

To differentiate a function with a single (independent) variable : $y = f(x)$ we write: $dy = f'(x)dx$. But if the independent variable x is a function of t, we write the function as $y = f(x(t))$ and to differentiate this we have to use the chain rule, and to do so, we will write the function as a composite then use the chain rule:

$$y = f(g(t)) = (f \circ g)(t).$$

Then differentiating once gives,

$$y' = f'(g(t)) \cdot g'(t) \qquad (6.10)$$

$$\text{Or,} \quad \frac{dy}{dt} = \frac{dy}{dx} \cdot \frac{dx}{dt}. \qquad (6.11)$$

Equation (6.11) is called Leibniz notation.

If $z = f(x, y)$, and $x = x(t)$, $y = y(t)$, and both are

differentiable then,

$$z = f(x,y) \tag{6.12}$$
$$\frac{dz}{dt} = \frac{\partial z}{\partial x}\frac{dx}{dt} + \frac{\partial z}{\partial y}\frac{dy}{dt} \tag{6.13}$$

For a smooth function $z = f(x,y)$, function of two variables,

$$dz = \frac{\partial z}{\partial x}dx + \frac{\partial z}{\partial y}dy, \tag{6.14}$$
$$dz = z_x dx + z_y dy. \tag{6.15}$$

Equation (6.15) is called the **total differential** of z.

And IF $z = f(x,y)$, $x = f(s,t)$, and $y = f(x,t)$, then this can be represented in tree diagram as follows:

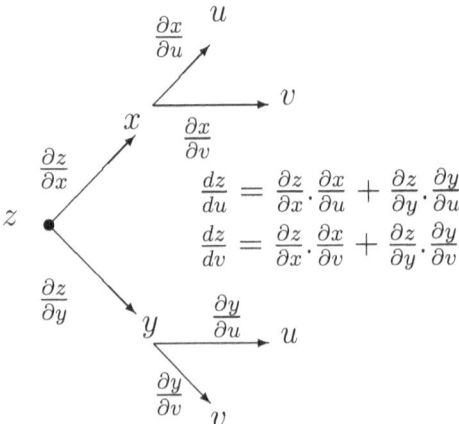

Now going back to the **Question** at the beginning of this section: Find the change in the volume of cylinder,

$$V = \pi r^2 h, \tag{6.16}$$

of radius r and height h when the radius is decreased by dr, and height increased by dh.

6.1. THE CHAIN RULE

Solution: Applying the chain-rule to differentiate equation (6.16) with respect to r and h, we get the total differential of the volume,

$$dV = \frac{\partial V}{\partial r} dr + \frac{\partial V}{\partial h} dh \qquad (6.17)$$
$$= 2\pi r h \, dr + \pi r^2 \, dh. \qquad (6.18)$$

Taking the initial volume to be V_0, the new volume V will be:

$$V = \pi(r - dr)^2(h + dh). \qquad (6.19)$$

Then we can compute the change in the volume as:

$$\Delta V = V - V_0$$
$$\text{Or, } \Delta V = \pi(r - dr)^2(h + dh) - \pi r^2 h.$$

Example- 5: For the given cylinder, let $r = 2$, $h = 8$, $dr = -.1$, and $dh = .2$, then $\Delta V = -7.54$ is the change in the volume of the cylinder.

For a smooth function with three independent variables, the total differential is:

$$w = f(x, y, z),$$
$$dw = \frac{\partial w}{\partial x} dx + \frac{\partial w}{\partial y} dy + \frac{\partial w}{\partial z} dz, \text{ or}$$
$$dw = w_x dx + w_y dy + w_z dz.$$

Example- 6:
Compute the total differential for the function:

$$u = x^2 + 2xy - y^3. \qquad (6.20)$$

Solution:
Differentiating equation (6.20) with respect of two independent variables x, and y, gives,

$$\begin{aligned} du &= u_x dx + u_y dy \\ &= (3x^2 + 2y)dx + (2x - 3y^2)dy. \end{aligned}$$

Differentiating Implicit Functions:

The Chain-Rule can be used to derive formula for Implicit Functions: Let,
$F(x, y) = 0$ with $x = x(t)$, and $y = y(t)$,
then $F(x(t), y(t)) = 0$ for all t in the domain.

Then differentiating with respect to t,

$$\frac{d}{dt} F(x(t), y(t)) = \frac{\partial F}{\partial x} \frac{dx}{dt} + \frac{\partial F}{\partial y} \frac{dy}{dt}. \tag{6.21}$$

If $x(t) = t$, then the above equation becomes,

$$\frac{d}{dt} F(x(t), y(t)) = \frac{\partial F}{\partial x} + \frac{\partial F}{\partial y} \frac{dy}{dt} = 0. \tag{6.22}$$

From this equation we get:

$$\frac{dy}{dt} = -\frac{\frac{\partial F}{\partial x}}{\frac{\partial F}{\partial y}} = -\frac{F_x}{F_y}, \quad f_y \neq 0. \tag{6.23}$$

Velocity and Acceleration:

Now, we can combine our study of parametric equations, curves, and vector valued functions to form a model for the motion along a curve, which is used in physics.
In the previous chapter we found the directional vector to be:

$$r(t) = x(t)i + y(t)j. \tag{6.24}$$

6.1. THE CHAIN RULE

Where both x, and y are functions of t. Differentiating (6.24) with respect t gives the velocity:

$$Velocity = v(t) = \frac{dr(t)}{dt} = \frac{dx}{dt}i + \frac{dy}{dt}j, \quad (6.25)$$
$$Or,\ v(t) = r'(t) = x'(t)i + y'(t)j. \quad (6.26)$$

And differentiating (6.26) with respect to t gives the acceleration:

$$Acceleration = a(t) = \frac{dv}{dt} = \frac{dx'}{dt}i + \frac{dy'}{dt}j,$$
$$Or,\ a(t) = v'(t) = r''(t) = x''(t)i + y''(t)j.$$

Then, by taking the norm of the velocity in equation (6.26) we can find the speed:

$$Speed = \|v(t)\| = \|r'(t)\| = \sqrt{(x'(t))^2 + (y'(t))^2}. \quad (6.27)$$

Example- 7:
A time dependent directional function describes the motion of a spider in a circular path, is given as:

$$r(t) = 3sin(5/2t)i + 3cos(5/2t)j. \quad (6.28)$$

Find the speed and acceleration of the spider in the path at any time t.

Solution:
The given equation has the following components:

$$x = 3sin(5/2)t$$
$$y = 3cos(5/2)t.$$

then, $x^2 + y^2 = (3sin(5/2)t)^2 + (3cos(5/2)t)^2 = 9$.
This gives the radius of the path to be $= 3$.

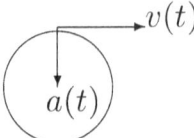

Taking any point on the path at any time the direction is:

$$\begin{aligned} Direction &= r(t) \\ &= 3\sin(5/2t)i + 3\cos(5/2t)j. \end{aligned}$$

$$\begin{aligned} Velocity &= v(t) = r'(t) \\ &= (15/2)\cos(5t/2)i - (15/2)\sin(5t/2)j. \end{aligned}$$

$$\begin{aligned} Acceleration &= a(t) = v'(t) \\ &= -(75/4)\sin(5t/2)i - (75/4)\cos(5t/2)j. \end{aligned}$$

$$\begin{aligned} Speed &= \|v(t)\| \\ &= \sqrt{(15/2\cos(5t/2))^2 + (15/2\sin(5t/2))^2} \\ &= 15/2 = 7.5. \end{aligned}$$

So the spider is moving around the curve in a circular path with the speed of 7.5 units.

6.2 Important Formula's

1. <u>Trigonometric Identities</u>

$$
\begin{aligned}
sin^2\theta + cos^2\theta &= 1. \\
1 + tan^2\theta &= sec^2\theta. \\
1 + cot^2\theta &= csc^2\theta. \\
sin(-\theta) &= -sin\theta. \\
cos(-\theta) &= cos\theta. \\
sin(2\theta) &= 2sin\theta\, cos\theta. \\
cos(2\theta) &= cos^2\theta - sin^2\theta. \\
&= 2cos^\theta - 1. \\
&= 1 - 2sin^2\theta. \\
cos^2\theta &= 1/2(1 + cos2\theta). \\
sin^2\theta &= 1/2(1 - cos2\theta).
\end{aligned}
$$

2. <u>Trigonometric Integrals</u>

$$
\begin{aligned}
\int \frac{1}{x^2 + b^2}dx &= \frac{1}{b}arctan(\frac{x}{b}).\ b \neq 0. \\
\int \frac{1}{x^2 - b^2}dx &= \frac{1}{2b}log(\frac{x-b}{x+b}). \\
\int \frac{1}{b^2 - x^2}dx &= \frac{1}{2b}log(\frac{x+b}{b-x}). \\
\int \frac{1}{\sqrt{x^2 + b^2}}dx &= log(x + \sqrt{x^2 + b^2}). \\
\int \frac{1}{\sqrt{x^2 - b^2}}dx &= log(x + \sqrt{x^2 - b^2}). \\
\int \frac{1}{\sqrt{b^2 - x^2}}dx &= arcsin\frac{x}{b}.
\end{aligned}
$$

3. Integration by Parts

$$\int xe^x dx = xe^x - e^x.$$

$$\int \log x \, dx = x \log x - x.$$

$$\int x \log x \, dx = 1/2 x^2 \log x - 1/4 x^2.$$

$$\int x \sin x \, dx = \sin x - x \cos x.$$

$$\int x \cos x \, dx = \cos x + x \sin x.$$

4. Other Integrals

$$\int \sin^2 x \, dx = 1/2(x - \sin x \cos x).$$

$$\int \cos^2 x \, dx = 1/2(x + \sin x \cos x).$$

$$\int \tan^2 x \, dx = \tan x - x.$$

$$\int \sec^2 x \, dx = 1/2 \sec x \tan x + 1/2 \ln|\sec x + \tan x|.$$

$$\int \sin^2 x \cos^2 x \, dx = 1/8 x - 1/32 \sin 4x.$$

$$\int \sec^3 x \, dx = 1/2 \sec x \tan x + 1/2 \log(\sec x + \tan x).$$

$$\int \frac{\sqrt{x^2 - b^2}}{x} dx = \sqrt{x^2 - b^2} - b \sec^{-1}|\frac{x}{b}|.$$

$$\int \sqrt{b^2 - x^2} \, dx = \frac{x}{2}\sqrt{b^2 - x^2} + \frac{x^2}{2} \sin^{-1}\frac{x}{b}.$$

5. Exponential and logarithmic Integrals

$$\int xe^{bx}dx = \frac{1}{b^2}(bx-1)e^{bx}.$$

$$\int (linx)^n dx = x(lnx)^n - n\int (lnx)^{n-1}dx.$$

$$\int \frac{(lnx)^n}{x}dx = \frac{(lnx)^{n+1}}{n+1}.$$

Index

A

Acceleration, 217
Application of Laplace, 167, 173
Application of 1st ODE, 60
Application of PDE, 210

B

Bernoulli, D., 58
Bernoulli's eq's, 58,
By inspection, 42, 105
By undetermined coef., 106
By superposition, 107

C

Cauchy Euler, 12, 119, 122
Cauchy-Riemann, 12
Chain rule, 213
Chapter-1 assignment, 30
Chapter-2 assignment, 81
Chapter-3 assignment, 139
Chapter-4 assignment, 183
Chapter-5 assignment, 207

Classification of DE, 12
Cramer's rule, 111

D

D'Alembert J., 95
Decomposition of PF, 151
Definition and Terminology of DE, 15
Differentiation of PS., 189
Direction formula, 124

E

Euler L., 12
Euler-Cauchy method, 12
Existence and uniqueness, 24, 92
Exact equations, 40, 43, 45
Explicit solutions, 19

F

Falling object, 33
Faraday's law, 167
Finding 2nd Soln, 93

Index

First ODE, 33

G

General remark, 18, 21

H

Historical notes, 11, 58
Homogeneous equations, 50, 51, 99

I

Index, 222
Implicit solutions, 20, 216
Important formula's, 219
Initial value problems, 22, 190
Integrating factor, 46
Inverse LT, 147

J

James Clark, 8
Jean Bernoulli, 12
Josiah Willard, 8

K

Kirtchhoff G., 167
Kirtchhoffs law, 167

L

Laplace differentiation, 142
Laplace inverse, 147
Laplace for unit-step, 160
Laplace transform, 141
Libiniz, 12
Linear DE, 16, 46, 47
Linear 2nd ODE, 83

M

Method of Red. of Order, 95, 97
Mixing problems, 60

N

Newton's classification, 11
Newton's law, 34
Newton's second law of motion, 34
Non-Homog. eq., 14, 52, 104
Non-linear eqs, 54, 55, 56,
Non-real, 103
Non-repeated, 152

O

Order of differential equations, 14
Ordinary DE, 13, 14

P

Partial Diff of chain rule, 209,

Particular Solution, 14, 52
Power series, 185
Property of LT, 146
Property of the Wronskian, 88

Q

Quadratic equation, 100
Quadratic factors, 154

R

RL-Circuit, 169
RLC-Circuit, 171
Real and Distinct, 101
Reduction of order, 102, 123
Repeated factors, 153
Rudolf Lipschitz, 12
Review exercise, 25, 66, 125, 179, 202
Review Ex. Soln., 27, 68, 128, 180, 202
Review from Calculus, 209

S

Separable eqs, 36
Solution of ODE, 148
Series solutions, 185
Step-size functions, 167

T

Taylor series, 192
Technical writing, 24, 60, 125, 179, 202
Total Derivative, 214
Trigonometric series, 191
Tree diagram, 214

U

Undetermined coefficients, 106
Unit step functions, 167

V

Variation of parameters, 109, 110, 112, 114
Velocity and Acceleration, 217

W

Wronskian, 87

Z

www.ingramcontent.com/pod-product-compliance
Lightning Source LLC
Chambersburg PA
CBHW030923180526
45163CB00002B/445